Assèta Kagambega

Caractérisation moléculaire de Salmonella et des pathovars de E. coli

AF185567

Assèta Kagambega

Caractérisation moléculaire de Salmonella et des pathovars de E. coli

Salmonella et Pathovars de E.coli sur la viande, les animaux de boucherie et la volaille

Presses Académiques Francophones

Impressum / Mentions légales

Bibliografische Information der Deutschen Nationalbibliothek: Die Deutsche Nationalbibliothek verzeichnet diese Publikation in der Deutschen Nationalbibliografie; detaillierte bibliografische Daten sind im Internet über http://dnb.d-nb.de abrufbar.

Alle in diesem Buch genannten Marken und Produktnamen unterliegen warenzeichen-, marken- oder patentrechtlichem Schutz bzw. sind Warenzeichen oder eingetragene Warenzeichen der jeweiligen Inhaber. Die Wiedergabe von Marken, Produktnamen, Gebrauchsnamen, Handelsnamen, Warenbezeichnungen u.s.w. in diesem Werk berechtigt auch ohne besondere Kennzeichnung nicht zu der Annahme, dass solche Namen im Sinne der Warenzeichen- und Markenschutzgesetzgebung als frei zu betrachten wären und daher von jedermann benutzt werden dürften.

Information bibliographique publiée par la Deutsche Nationalbibliothek: La Deutsche Nationalbibliothek inscrit cette publication à la Deutsche Nationalbibliografie; des données bibliographiques détaillées sont disponibles sur internet à l'adresse http://dnb.d-nb.de.

Toutes marques et noms de produits mentionnés dans ce livre demeurent sous la protection des marques, des marques déposées et des brevets, et sont des marques ou des marques déposées de leurs détenteurs respectifs. L'utilisation des marques, noms de produits, noms communs, noms commerciaux, descriptions de produits, etc, même sans qu'ils soient mentionnés de façon particulière dans ce livre ne signifie en aucune façon que ces noms peuvent être utilisés sans restriction à l'égard de la législation pour la protection des marques et des marques déposées et pourraient donc être utilisés par quiconque.

Coverbild / Photo de couverture: www.ingimage.com

Verlag / Editeur:
Presses Académiques Francophones
ist ein Imprint der / est une marque déposée de
AV Akademikerverlag GmbH & Co. KG
Heinrich-Böcking-Str. 6-8, 66121 Saarbrücken, Deutschland / Allemagne
Email: info@presses-academiques.com

Herstellung: siehe letzte Seite /
Impression: voir la dernière page
ISBN: 978-3-8416-2113-9

UNIVERSITE DE OUAGADOUGOU

École Doctorale
Sciences et Technologies

Laboratoire : Centre de Recherche en Sciences
Biologiques Alimentaires et Nutritionnelles
(CRSBAN)

N° d'ordre............/

THÈSE

Présentée par

Assèta KAGAMBEGA épouse BOUGMA

Pour obtenir le grade de

Docteur de l'Université de Ouagadougou

Option: Sciences Biologiques Appliquées

Spécialité: Biochimie-Microbiologie

Caractérisation phénotypique et moléculaire de souches de *Salmonella* spp et de *Escherichia coli* isolées des viandes et des fèces d'animaux de boucherie et de volaille au Burkina Faso

Soutenue le 29 Décembre 2012 devant le Jury composé de:

Président: **Pr Alfred S. TRAORE**   *(Professeur Titulaire, Université de Ouaga)*

Membres: **Pr Nicolas BARRO**   *(Professeur Titulaire, Université de Ouaga)*

        Dr Kaisa HAUKKA   *(Professeur associé, Université de Helsinki)*

        Pr Idrissa SANOU   *(Professeur Agrégé, Université de Ouaga)*

        Dr Hagrétou SAWADOGO *(Maitre de Recherche, CNSRT, DTA)*

i

DEDICACES

Je dédie ce présent travail :

A mon défunt grand-père et à ma grand-mère qui m'ont élevée;

A mon père et à ma mère, pour la confiance qu'ils ont placée en moi en me soutenant jours et nuits;

A ma grande sœur Aminata pour son soutien immense et ses nombreux conseils;

A mes frères et soeurs pour leur soutien et leurs encouragements;

A ma tante Hadja Mariam et son fils El Hadj Souleymane pour leurs encouragements;

A Moussa Bougma, dont l'amour et la compréhension me soutiennent chaque jour!

A mon bébé Mariam Sarah Princesse pour sa douceur et la joie immense qu'elle m'apporte a chaque instant.

A toute ma famille pour leur encouragement.

REMERCIEMENTS

Ce travail a pu être réalisé grâce au soutien d'institutions et aux conseils de nombreuses personnes; Il est également le fruit d'une collaboration internationale entre le CRSBAN (Burkina Faso) et le Laboratoire national de santé public et du bien être (THL) de la Finlande dans le cadre du projet « *Epidemiology of water- and foodborne bacterial infections in Burkina Faso (West Africa) - strategies for sustainable national public health management* ».

Qu'il nous soit permis de remercier:
Les autorités de l'Université de Ouagadougou à qui nous exprimons toute notre reconnaissance.

Le Pr Alfred S. TRAORE, Directeur du Centre de Recherche en Sciences Biologiques Alimentaires et Nutritionnelles (CRSBAN), Responsable Académique des formations Doctorales en Biochimie et Microbiologie/Biotechnologies de l'Université de Ouagadougou, Président du Réseau Ouest Africain de Biotechnologies (RA-Biotech). Toute notre reconnaissance pour nous avoir acceptés dans son Centre et de nous avoir soutenus sur ce thème qui a un intérêt capital.

Le Pr Nicolas BARRO, Pr Titulaire, Responsable du Laboratoire de Biochimie et de Biologie Moléculaire, d'épidémiologie et de surveillance des Bactéries et virus transmis par les aliments /CRSBAN, Directeur de cette thèse qui malgré ses multiples occupations a suivi de bout à bout ce travail depuis le DEA sur le sujet. Merci énormément pour le suivi quotidien, les nombreux conseils et toute l'attention accordée à ce travail. Sans oublier votre soutien immense sur le plan social, recevez toute notre profonde gratitude. Merci encore et encore Professeur!

Le Dr Kaisa HAUKKA, chercheur à l'institut National de santé publique et de bien être de Finlande, coordonnatrice du projet, *pour avoir suivi au quotidien ce travail, son devouement total à ce travail et pour toute sa disponibilité.*

Le Pr Anja SIITONEN, directrice du laboratoire de microbiologie de l'institut national de santé publique et de bien être de Finlande, pour son soutien technique et pour nous avoir acceptés dans son laboratoire.

Nos remerciements vont à tous les rapporteurs de cette thèse
Au Pr Amy Gassama SOW de l'Institut Pasteur de Dakar pour ses énormes contributions formulées lors de l'évaluation de la thèse;
Au Pr Idrissa SANOU, du Laboratoire de Bactériologie Virologie du CHU-YO, tous nos sincères rémerciements pour l'évaluation et les conseils d'amélioration de cette thèse;
Au Dr Hagretou LINGANI-SAWADOGO, Maître de recherche au CNRST/IRSAT/DTA, vous êtes un modèle de femmes scientifiques. Nous vous remercions pour l'évaluation de cette thèse et pour les conseils formulés pour nous dans ce monde de femmes chercheurs.
La Fondation Internationale pour la Science (IFS) pour avoir financé une partie de ce travail

Nous remercions le Groupe AGISAR-WHO et sa coordonnatrice Dr Awa AIDARA-KANE pour le soutien financier apporté à notre équipe de recherche lors de la finalisation de cette thèse de doctorat Unique.

Nos remerciements vont à tous les Enseignant-Chercheurs du CRSBAN de l'Université de Ouagadougou :
Pr Aboubakar S. OUATTARA, Pr Yves TRAORE, Dr Philippe A. NIKIEMA, Dr Cheik A. T. OUATTARA, Dr André Jules ILBOUDO, Dr Marcel D. BENGALY, Dr Aly

SAVADOGO, Dr Imaël N. BASSOLE pour l'encadrement pédagogique que nous avons reçu et aussi pour leurs conseils et leurs encouragements.

A tous nos Maîtres formateurs du RA-BIOTECH pour les cours et les conseils reçus tout au long de la formation de qualité que vous assurez.

Nous remercions :
Le personnel du l'abattoir frigorifique de Ouagadougou et les différents bouchers pour leur franche collaboration.

Le doctorant Taru LIENMAN, Outi MARTIKAINEN pour leur soutien et les bons moments passés ensemble.

Nous remercions tout le personnel de l'unité de bactériologie de l'institut National de Santé publique et du bien être (THL) de Finlande pour leur assistance et leur bonne humeur. Ce fut un plaisir de travailler avec vous!

Nous remercions tous les doctorants pour leur soutien: Cheikna ZONGO, Joseph SAWADOGO, Léon NITIEMA, Zénabou SEMDÉ, Oumar TRAORE, Gertrude TCHAMBA, Réné DEMBÉLÉ, Saidou KABORÉ, Larissa WANRÉ, Adissa OUEDRAOGO, Rakieta KOURAOGO, Nicolas OUEDRAOGO, Fousseni YARO, Kuan TRAORÉ, Nafissatou OUEDRAOGO. Merci et surtout bon courage à vous pour la suite de vos travaux.
A tous mes camarades de classe de la 6ème promotion du RA-BIOTECH pour les moments passés ensemble.
A tous mes amis pour leurs encouragements et leur soutien. Nos remerciements vont à toute ma famille et à ma belle famille pour le soutien au quotidien.
Merci à tous ceux qui ont contribué de près ou de loin à la réalisation de ce travail.

SOMMAIRE

Liste des articles et posters

A- Articles

1. **Kagambèga A.,** Haukka K., Siitonen A., Traore A. S., Barro N. 2011. Prevalence of *Salmonella enterica* and the hygienic indicator *Escherichia coli* in raw meat at markets in Ouagadougou, Burkina Faso. *J. Food Prot.* 74, 1547–1551.

2. **Kagambèga A.,** Martikainen O., Lienemann T., Siitonen A., Traoré A. S., Barro N., K. Haukka. 2012. Diarrheagenic *Escherichia coli* detected by 16-plex PCR in beef intestine and raw meat sold in Ouagadougou, Burkina Faso. *Int. J Food Microbiol.* **153**, 154–158.

3. **Kagambèga A.,** Barro N., Traoré A.S., Siitonen A., Haukka K., 2012. Characterization of *Salmonella enterica* and detection of the virulence genes specific to diarrheagenic *Escherichia coli* from poultry carcasses in Ouagadougou, Burkina Faso. *Foodborne pathog. Dis.* DOI: 10.1089/fpd.2011.1071).

4. **Kagambèga A.,** Martikainen O., Siitonen A., Traoré A. S., Barro N., Haukka K. 2011. Prevalence of diarrheagenic *Escherichia coli* in the feces of slaughtered animals in Burkina Faso. (MicrobiologyOpen. Doi: 10.1002/mbo3.30).

5. **Kagambèga A.,** Lienemann T., Siitonen A., Traoré A. S., Barro N., Haukka K. 2012. Antimicrobial susceptibility, PFGE and serovars of *Salmonella* from cattle, poultry, pigs, hedgehogs and humans in Burkina Faso. (A soumettre pour publication).

6. Martikainen O., **Kagambèga A.,** Bonkoungou I.J., Barro N., Siitonen A., Haukka K., 2012. Characterization of Shigatoxigenic *Escherichia coli* strains from Burkina Faso (Sous presse, *Foodborne pathog. Dis*).

B- Communications orales et affichées

1. Haukka K., **Kagambèga A.**, Martikainen O., Traore A. S., Barro N., Siitonen A. 2011. Common occurrence of various STEC serotypes in meats sold at local markets in Ouagadougou, Burkina Faso. Poster. FEMS, 2011, 4th congress of European Microbiologists, Geneva, Switzerland, June 26-30.

2. Haukka K., **Kagambèga A.**, Martikainen O., Barro N., Traore A.S., Siitonen A. 2010. Detection and characterization of EHEC isolates from meat samples collected from local markets in Ouagadougou, Burkina Faso. Oral communication, 9th International Meeting on Microbial Epidemiological Markers (IMMEM-9) in Wernigerode, Germany, September 1-4.

3. **Kagambèga A.**, Barro N., Kuusi M., Haukka K., Siitonen A., Traoré A.S. 2011. Prevalence of some enteropathogenic bacteria on raw meat and poultry from tree popular markets of Ouagadougou, Burkina Faso. Poster. Journées Portes ouvertes sur les activités des Doctorants, 2011, Université de Ouagadougou.

Liste des abréviations

AGISAR: Advisory Group on Integrated Surveillance of Antimicrobial Resistance

CLED: Cystine-Lactose Electrolyte-Deficient

CRSBAN: Centre de Recherche en Sciences Biologiques, Alimentaires et Nutritionnelles

EAEC: *Escherichia coli* entéroaggrégatif

EFSA: European Food Safety Autority

EHEC: *Escherichia coli* entérohémorragique

EIEC: *Escherichia coli* entéroinvasif

EMB: Eosin Methylene Blue agar

EPEC: *Escherichia coli* entéropathogène

EPT: Eau Peptonee Tamponée

ETEC: *Escherichia coli* entérotoxinogène

FAO : Organisation des Nations-Unies pour l'alimentation et l'agriculture

FNUAP: Fonds des Nations Unies pour la Population

HACCP: Hazard Analyzis Critical Control Point

INSD: Institut National de la Statistique et de la Démographie

ONU: Organisation des Nations Unies

OMS : Organisation mondiale pour la santé

PFGE: Electrophorèse en champs pulsé (Pulse-field gel electrophoresis)

PTT: Purpura Thrombotique Thrombocytopénique

g : gravité

g: gramme

ml: millilitre

µl: microlitre

µg: microgramme

SHU: Syndrome Hémolytique et Urémique

SMAC: Sorbitol MacConkey Agar

STEC: *Escherichia coli* productrice de Sigha-toxine

UEMOA: Union Economique et Monétaire Ouest Africaine

XLD: Xylose Lysine Desoxycholate Agar

H-buffer: Tampon H

SOGEAO: Société de Gestion des Abattoir de Ouagadougou

Résumé

Les maladies d'origine alimentaire posent un problème de santé publique à travers le monde, et plus particulièrement dans les pays en développement dont le Burkina Faso, où l'hygiène alimentaire reste un défi à relever. Les agents principaux en cause dans ces maladies sont les bactéries surtout les Salmonelles et les souches pathogènes de *Escherichia coli*. En effet, tous les ans, des milliers de personnes sont victimes de maladies d'origine alimentaire qui sont généralement associées à la consommation de produits carnés notamment le boeuf, le poulet et le porc. Malheureusement, au Burkina Faso, très peu de données existent sur la prévalence, l'épidémiologie et l'écologie de ces bactéries.

La première partie du présent travail, s'est consacrée à l'isolement de *Salmonella* et des pathovars d'*E. coli* dans des échantillons de viande de diverses origines (bœufs, moutons, poulets). L'étude a été réalisée sur 150 échantillons de viandes dans 4 marchés de la ville de Ouagadougou, capitale du Burkina Faso.

La seconde partie de ce travail a porté sur la recherche de *Salmonella* d'une part dans 729 échantillons de fèces d'animaux de boucherie prélevés au cours de l'abattage (304 fecès de bœufs, 50 de porcs, 350 poulets et 25 de hérisson) ; et d'autre part, des pathovars de *E. coli* dans 704 échantillons de fèces d'animaux (304 fecès de bœuf, 50 de porcs et 350 de poulets).

La recherche de la relation épidémiologique entre les différentes souches (souches isolées des viandes, animaux et des patients malades) a été réalisée par l'étude des différents marqueurs phénotypiques telque le sérotypage et la sensibilé aux antibiotiques. L'analyse génotypique a été réalisée par différentes techniques telle que la PCR multiplexe et le profil de l'ADN total par électrophorèse en champ pulsé (PFGE).

Les résultats de la contamination par *Salmonella* et des pathovars de *E. coli* des viandes ont montré des pourcentages de 9,3% et 43% respectivement, alors que pour les fèces, la prévalence de *Salmonella* et pathovars de *E. coli* est de 53% et 59% respectivement. L'analyse sérotypique des *Salmonella* d'origine carnée et fécale a révélé une diversité de sérotype. Les souches de *Salmonella* des viandes ont montré une résistance faible vis-à-vis de douze antibiotiques testé, alors que celles des fèces d'animaux et des patients malades ont montré une résistance relativement élevée avec des multi-résistances associant deux à cinq classes d'antibiotique chez certains sérotypes comme *S.* Hato, *S.* Ruiru, *S.* Typhimurium, *S.* Urbana et *S.* Virchow. La combinaison des différents résultats, aussi bien de l'approche phénotypique que génotypique, suggère que les souches de *Salmonella* d'origine humaine et alimentaire présentent une grande similitude et une diversité remarquable.

La PCR multiplexe des *E. coli* a montré une prédominance des *E. coli* producteurs de Shiga-toxine (STEC) aussi bien au niveau des viandes qu'au niveau des fèces d'animaux.

Les marqueurs phénotypiques ont montré des résultats importants pour les premières étapes d'investigations épidémiologiques des infections à *Salmonella* et *E. coli*. Les marqueurs moléculaires essentiellement l'électrophorèse en champ pulsé permettent d'obtenir une discrimination plus précise entre les souches.

Au meilleur de notre connaissance, c'est la première étude sur la caractérisation phénotypique et moléculaire des salmonelles et pathovars de *E. coli* d'origine humaine, alimentaire et animale au Burkina Faso.

Mots clés : Épidémiologie, *Salmonella*, *E. coli*, PCR, PFGE, Viande, animaux, Burkina Faso.

Abstract

Food-borne illnesses are a public health concern worldwide and are more pronounced in developing countries particularly in Burkina Faso, due to the poor hygienic conditions. Most of the food-borne enteric infections are caused by *Salmonella* and diarrheagenic *Escherichia coli*. Every year, many people are suffering from foodborne diseases generally linked to the consumption of meat and meat product, especially beef, poultry and pork. Unfortunately, in Burkina Faso, little data are available on the occurrence of *Salmonella* and diarrheagenic *E. coli* in food of animal origin or animals.

The first part of the present study was on the isolation of *Salmonella* and the pathogroup of diarrheagenic *E. coli*. A total of 150 samples of beef meat, beef intestine, mutton and chicken were collected from four local markets for investigation. The first part of this work has been dedicated to the isolation of *Salmonella* and the pathogroups of *E. coli* in meat samples from various sources (cattle, sheep, chickens). The study was performed in 150 meat samples from four markets of Ouagadougou.

The second part of this work has focused first, on the detection of *Salmonella* in 729 fecal samples of slaughtered animals and African pigmy hedgehog (304 feces of cattle, 50 pigs, 350 chickens and 25 of hedgehog), and secondly, on the detection of the pathogroups of diarrheagenic *E. coli* in 704 slaughtered animals feces (304 feces of cattle, 350 of chickens and 50 of pigs). The epidemiological relationship between strains (strains isolated from meat, animals and patients samples) was carried out by the phenotypic markers detection using serotyping and the antibiotic susceptibility. Genotypic analysis was performed by different techniques such as multiplex PCR and pulsed field gel electrophoresis (PFGE).

The *Salmonella* and diarrheagenic *E. coli* contamination in meat was 9.3% and 43% respectively, while from the feces, the prevalence of these pathogens was 53% and 49% respectively. Analysis of *Salmonella* serotypes isolated from meat and fecal origin revealed a diversity of serotype. *Salmonella* strains isolated from meat has shown a low resistance to twelve antibiotics, whereas the strains from animals feces and patients have shown a high resistance to one or more antibiotics; for example resistance to two or five antibiotics was observed from *S.* Hato, *S.* Ruiru, *S.* Typhimurium, *S.* Urbana and *S.* Virchow.

The combination of the different results, phenotypic and genotypic approach, suggests that *Salmonella* strains from food of animal origin, animals or human have a high similarity and a remarkable diversity. The multiplex PCR of diarrheagenic *E. coli* showed a predominance of Shiga-toxin producing *E. coli* (STEC) from meat and animal feces samples.

The phenotypic markers showed significant results for the early stages of the epidemiological investigations of *Salmonella* and diarrheagenic *E. coli*. Molecular markers, essentially, PFGE enable a more accurate discrimination between strains. To our knowledge, this is the first study of phenotypic and molecular characterization of *Salmonella* and diarrheagenic *E. coli* from food of animal origin, animals and human in Burkina Faso.

Keywords: Epidemiology, *Salmonella*, diarrheagenic *E. coli*, PCR, PFGE, meat, animals, Burkina Faso

Introduction

Les maladies infectieuses transmises par les aliments constituent un problème de santé publique et économique important dans le monde (White *et al.*, 1997). Les risques d'infection des populations sont de plus en plus élevés et sont liés aux nouvelles pratiques d'alimentation, aux changements en production animale, aux conditions socio-économiques et aux demandes accrues du consommateur.

Cependant, il est difficile d'obtenir des évaluations précises de l'incidence des maladies infectieuses transmises par les aliments. Par exemple dans les pays développés, on a rapporté que le pourcentage annuel des personnes souffrant de maladies infectieuses transmises par les aliments atteint 30%, alors que même s'il y est moins bien documenté, le gros du problème pèse sur les pays en développement (OMS, 2007). La forte prévalence des affections diarrhéiques dans nombre de ces pays est le signe de problèmes sous-jacents importants de salubrité des aliments.

Plusieurs agents sont responsables de ces maladies parmi lesquels, trois principales bactéries pathogènes à savoir *Campylobacter*, *Salmonella* et *Escherichia coli* sont les plus dominantes à travers le monde (Meng *et al.*, 1998). Les salmonelloses et les diarrhées causées par les *E. coli* sont les plus présentes en Afrique et plus particulièrement au Burkina Faso où les maladies diarrhéiques constituent la principale cause de morbidité et de mortalité surtout chez les enfants (Barro *et al.*, 2005; Simporé *et al.*, 2009).

Les salmonelloses provoquent des symptômes d'une large gamme de sévérité, allant de légers maux de ventre à des degrés divers d'entérites, jusqu'aux septicémies et dans les cas extrêmes la mort. Les salmonelles non typhoïdiques constituent la grande majorité des isolats appartenant au genre *Salmonella* et représentent la principale cause de gastro-entérites dans le monde (Humphrey, 2000; Molla *et al.*, 2003). Les pathovars de *E. coli* sont généralement classés en plusieurs groupes en fonction de leurs gènes de virulence. Les 5 principaux pathovars sont: *E. coli* entéropathogène (EPEC), *E. coli* entérotoxinogène (ETEC), *E. coli* productrice de Shiga-toxine (STEC), *E. coli* entéroinvasive (EIEC) et *E. coli* enteroaggrégative (EAEC) (Nataro et Kaper, 1998). Ces pathovars provoquent des pathologies

1

intestinales qui peuvent aller de la simple diarrhée à des complications graves chez l'homme. Par exemple, STEC est responsable du syndrome hémolytique et urémique (SHU) ou le purpura thrombotique thrombocytopénique (PTT) dont le pronostic peut être fatal (Rhoades *et al.*, 2009).

La contamination par les salmonelles et les pathovars de *E. coli* résulte principalement de l'ingestion de produits alimentaires contaminés; les viandes et les produits carnés étant les plus incriminés. En effet, ces produits sont susceptibles d'être souillés par la matière fécale de divers animaux (bœuf, porc, poulet, mouton…) dont le tube digestif apparaît comme le principal réservoir naturel de ces bactéries. Au Burkina Faso, l'élevage connait depuis 2000 un développement notable, mais les pratiques d'élevage et d'abattage accusent un retard technologique considérable par rapport aux pays industrialisés. Ceci retentissant non seulement sur la productivité, mais aussi et surtout sur la santé publique.

L'environnement joue un rôle important en santé publique, les germes s'y adaptent en développant des formes de résistance et aujourd'hui la résistance aux antibiotiques des espèces pathogènes est un problème scientifique et de développement à résoudre à l'échelle internationale. Ces résistances aux antibiotiques constituent un autre aspect du problème de santé publique dû aux salmonelles et *E. coli*. Il est admis actuellement dans les pays développés mais aussi dans les pays en développement, qu'une partie des *Salmonella* ou des *E. coli* multi-résistantes retrouvées chez l'homme sont d'origine animale et ont acquis leurs gènes de résistance en élevage avant de les transmettre aux humains à travers les aliments (Ungemach *et al.*, 2006).

Une étude menée par Barro et Traoré, (2001) a montré une prévalence élevée des *E. coli* pathogènes et des Salmonelles au niveau des services de santé burkinabé, alors que leurs mécanismes de contamination des aliments et d'infection des hommes sont peu connus. Pour le contrôle des maladies infectieuses transmises par les aliments, des données sur la prévalence et la caractérisation de ces pathogènes sont nécessaires. Nous apportons notre contribution à cette nécessité par le présent travail réalisé sur le thème suivant : « **Caractérisation phénotypique et moléculaire de souches de**

Salmonella spp et de *Escherichia coli* isolées des viandes et des fèces d'animaux de boucherie et de volaille au Burkina Faso».

Notre étude a visé les objectifs suivants:

Objectif général

L'objectif général de ce travail est de contribuer à une meilleure connaissance des réservoirs de *Salmonella* et des pathovars de *Escherichia coli* ainsi que le mode de contamination de ces pathogènes à l'homme.

Objectifs spécifiques

- Déterminer la prévalence des Salmonelles et des pathovars de *E. coli* sur les viandes de grande consommation au Burkina Faso;
- Inventorier les principaux pathovars de *E. coli* et les principaux sérotypes de *Salmonella* circulant au Burkina Faso;
- Elucider les caractéristiques moléculaires et les aspects épidémiologiques (réservoir et mode de propagation) de ces bactéries au Burkina Faso;
- Evaluer la résistance aux antibiotiques des sérotypes de *Salmonella* isolés.

PREMIERE PARTIE:

REVUE BIBLIOGRAPHIQUE

Chapitre 1: Elevage, production de viande et sécurité sanitaire des aliments

I. Systèmes d'élevage et production de viande en Afrique de l'Ouest

L'élevage est une activité ancienne en Afrique. Il est particulièrement important pour les pays sahéliens qui exportent leurs animaux vers les pays côtiers comme le Nigeria ou la Côte d'Ivoire. Contrairement à l'Afrique de l'Est et du Sud, il existe peu de ranchs clôturés, d'élevages industriels type poulets ou porcs en batterie, ou de fermes laitières en Afrique de l'Ouest. L'élevage demeure donc "traditionnel"(Drabo *et al.*, 2001). Les systèmes d'élevage traditionnels sont représentés par:

- **l'élevage en région urbaine:** il concerne principalement la production laitière ainsi que l'embouche d'animaux domestiques provenant des zones rurales et il est très pratiqué en temps de crise économique. Dans ce type d'élevage, les ruminants ne sont pas toujours menés au pâturage.

- **le petit élevage rural:** ce type d'élévage est pratiqué depuis les zones humides jusque dans les régions désertiques, au niveau des oasis. Il concerne surtout les volailles, les petits ruminants (chèvres et moutons), mais parfois les bovins et les ânes, dans les zones arides. C'est un élevage de type sédentaire, avec peu d'intrants externes.

- **l'élevage agro-pastoral:** les bovins sont généralement nombreux ainsi que les dromadaires dans les zones arides. Les animaux divaguent librement ou sont au contraire gardés par un berger.

- **l'élevage pastoral:** l'élevage constitue l'unique ou la principale source de revenus. Le bétail comprend des bovins, des moutons, des chèvres et des dromadaires dans les régions les plus arides. Les pâturages correspondent à des zones impropres aux cultures. Cependant, les animaux broutent également les résidus de moissons après les récoltes. La mobilité est la principale caractéristique de ce type d'élevage.

I.1. Production de viande dans la sous région

En 2001 et 2002, la production de viande dans l'espace de l'Union Economique et Monétaire Ouest Africaine (UEMOA) se situait autour de 320000 tonnes pour la viande bovine (soit 0,6% de la production mondiale) et de 210000 tonnes pour les viandes ovine et caprine. Les importations de viande sont très faibles et ne concernent que quelques dizaines de tonnes. Les échanges intracommunautaires constituent l'essentiel du commerce de bétail. Ces échanges concernent principalement le bétail sur pied: la filière bétail-viande est la plus représentative des complémentarités existant entre les différents pays de l'UEMOA. Depuis le milieu des années 1980, le Mali, premier producteur, assure plus de 50% de la production dans cet espace régional tant pour les bovins que pour les caprins. Le Burkina Faso et le Sénégal en représentent chacun environ 16%, suivis par la Côte d'Ivoire et le Niger avec 13%.

I.2. Production de viande au Burkina Faso

Pays sahélien de 274 000 Km^2, le Burkina Faso est géographiquement enclavé et largement tributaire des aléas climatiques. L'élevage constitue la deuxième ressource du secteur primaire burkinabé, représentant 27,2% de sa valeur ajoutée: il s'agit d'un élevage extensif dont les actuels résultats paraissent très éloignés des potentialités réelles du pays. Bien qu'essentiellement traditionnel, l'élevage présente des opportunités importantes de développement dans le cadre des échanges sous régionaux. En 2005, le cheptel était estimé à 7,6 millions de têtes de bovins, 17,7 millions d'ovins et de caprins et 32 millions de volailles. Le cheptel bovin burkinabé se place au second rang des pays de l'UEMOA en importance, derrière celui du Mali (Service des statistiques animales, 2007).

Le pays est doté de trois abattoirs frigorifiques, 45 abattoirs séchoirs de brousse et des aires d'abattage dans la plupart des chefs lieu de département. La réhabilitation des installations de Ouagadougou en novembre 2004 a permis une réorganisation complète de ce secteur. Sous la tutelle d'une société de gestion, SOGEAO, les abattoirs fonctionnement aujourd'hui à plein régime. La production de viande pour l'année 2005 s'est élevée à 10846 tonnes pour une capacité de traitement de 11 000 tonnes. La transformation à des fins de boucherie ou de charcuterie reste peu développée et limitée aux boucheries spécialisées de Ouagadougou et Bobo-Dioulasso.

Dans la sous région ouest africaine, les conditions de production de carcasses restent un défi à relever. Au Burkina Faso, les carcasses sont produites dans des abattoirs dont la plupart ne répondent pas aux normes internationales. Les mesures d'hygiène ne sont pas respectées dans les abattoirs car la plupart des employés chargés de l'abattage n'ont pas la notion d'hygiène. En plus l'inspection post mortem est seulement basée sur l'observation macroscopique de l'aspect des carcasses, pourtant les bactéries pathogènes ne peuvent pas être détectées lors de ces observations. Le transport des viandes vers les marchés publiques se fait par différents moyens allant des bicyclettes à des véhicules non frigorifiés. L'exposition lors de la vente se fait également à température ambiante. Ces mauvaises conditions favorisent le développement de bactéries pathogènes.

Une surveillance des conditions d'hygiènes et un respect de normes internationales au niveau des abattoirs permettront d'exporter les viandes, ce qui contribuera énormement à l'expansion économique du Burkina Faso. Les différentes formes d'exposition de la viande au Burkina Faso sont representées en photos ci-dessous (Kagambèga *et al.*, 2011).

Photos : Conditions de traitement des viandes : A et B: Transports de carcasses; C et D: Conditions d'exposition lors de la vente des viandes; E: Plumaison de volaille; F: Entrepossage de carcasses de poulet (Photos: Kagambèga A.)

II. Sécurité sanitaire des aliments

La sécurité sanitaire des aliments tient compte de tous les risques chroniques ou aigus, susceptibles de rendre les aliments préjudiciables à la santé du consommateur (FAO/OMS, 2004). Les aliments peuvent être contaminés par des substances chimiques et/ou biologiques susceptibles de nuire à la santé du consommateur.

II.1. Contamination des aliments par des substances chimiques

Les dangers chimiques figurent parmi les principales causes de maladies d'origine alimentaire, bien que les effets soient souvent difficilement associables à un élément particulier et puissent intervenir longtemps après sa consommation (Randell, 2002). Plus récemment, la contamination par les dioxines présentes dans les aliments pour animaux a souligné conjointement l'importance du contrôle de toute la chaîne alimentaire et la prise de conscience au niveau international des problèmes posés par la sécurité sanitaire des aliments (FAO/OMS, 2004).

II.2. Contamination des aliments par des microorganismes pathogènes

Au cours des dernières décennies, une série de microorganismes est apparue comme responsables de maladies d'origine alimentaire. Différents types de protozoaires et de virus peuvent contaminer les aliments, par exemple *Cryptosporidium parvum*, *Toxoplasma gondii*, *Clonorchis sisensis*, virus Norwalk et hepatite A (King *et al.*, 2000; Wihelmi *et al.*, 2003). Plusieurs bactéries ont été identifiées en tant que cause importante de maladies d'origine alimentaire, par exemple, *Campylobacter jejuni*, *Vibrio parahaemolyticus* et *Yersinia enterocolitica* (Mead *et al.*, 1999). Compte tenu de la capacité d'adaptation des microorganismes, la modification des modes de production, de conservation et de conditionnement des produits alimentaires s'est traduite par une évolution des dangers en termes de salubrité des aliments.

Les graves épidémies dues aux bactéries pathogènes notamment *E. coli* et *Salmonella* ont mis en évidence des problèmes de sécurité sanitaire des aliments et une aggravation de l'inquiétude de la population face au risque de garanties insuffisantes en matière de santé publique offertes par les systèmes modernes de production agricole, de transformation des aliments et de commercialisation (Meng *et al.*, 1998; Todd, 1997). En dépit des progrès de notre connaissance des caractéristiques écologiques des organismes responsables des intoxications alimentaires et des conditions dans lesquelles ils risquent de se développer et de survivre, les moyens dont nous disposons pour éliminer certains d'entre eux sont plus limités. Cela tient sans doute en partie à la transformation des pratiques de production, à la maîtrise insuffisante des risques au niveau de l'exploitation agricole, aux difficultés de l'industrie à maîtriser les risques au cours de la production, à la demande croissante de produits frais, à l'évolution dans le sens d'un traitement réduit au minimum des denrées alimentaires et à l'allongement de la durée de stockage de nombreux aliments. Par exemple, *Salmonella* reste une cause importante d'intoxication alimentaire et son incidence progresse (Humphrey, 2000; Molla *et al.*, 2003). Largement répandue dans les troupeaux de bovins, *Salmonella* Typhimurium DT 104 résiste à plusieurs antibiotiques. Or, l'incidence de cet organisme progresse ainsi que le nombre d'isolats résistants aux antibiotiques. Plus d'un tiers des personnes contaminées par cet organisme doivent être hospitalisées, trois pour cent des cas étant mortels (vanden Bogaard et Stobberingh, 2000).

Le danger notable présenté par l'agent pathogène entérohémorragique *Escherichia coli* O157:H7 s'est largement manifesté par des épidémies majeures de cette maladie. Le caractère pathogène de *E. coli* O157:H7 a été identifié pour la première fois en 1982, mais l'absence de méthode de détection sensible a compliqué l'identification des réservoirs et des sources de cet organisme.

L'efficacité de la prévention et la lutte contre ces organismes exigent un effort pédagogique à grande échelle et le cas échéant, de nouvelles initiatives telles l'introduction des systèmes HACCP (Hazard Analysis Critical Control Point) au niveau de la production primaire.

III. Mécanismes de contamination des denrées alimentaires

Les aliments sont aujourd'hui des vecteurs de transmission de divers contaminants dont les microorganismes pathogènes et leurs métabolites. Ces microorganismes sont d'origine endogène et exogène. En effet, plusieurs études ont montré que des microorganismes de contamination des aliments notamment les viandes peuvent provenir de la matière première ou de son environnement (Leyral, 1996; Barro *et al.*, 2007). Barro *et al.* (2007) ont montré que les phénomènes de contamination des aliments sont beaucoup plus complexes (figure 1).

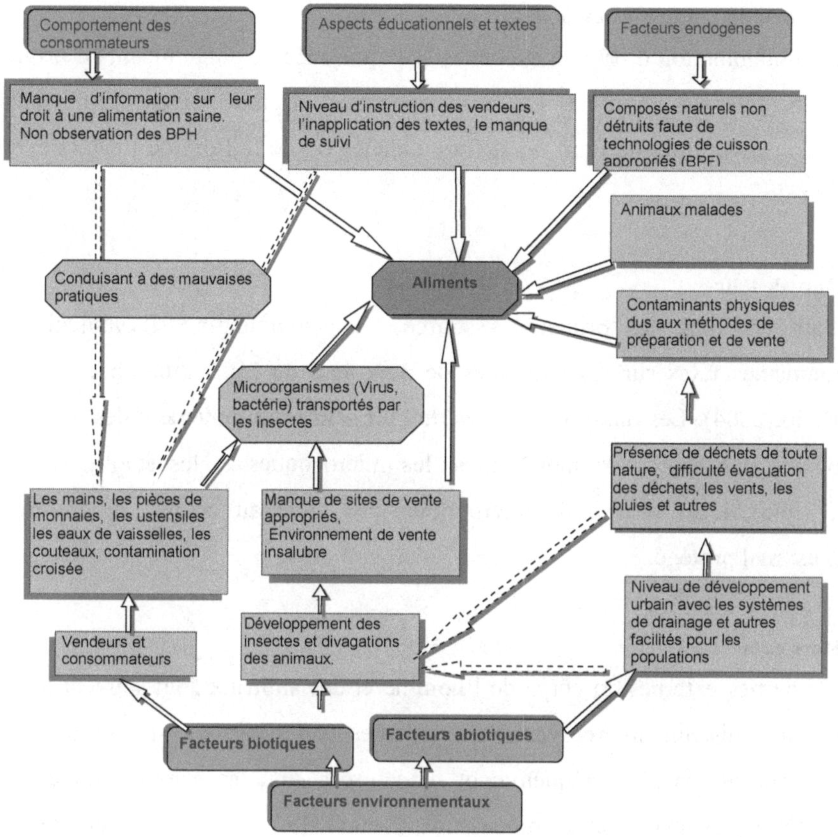

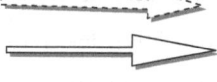

Figure 1: Diagramme des mécanismes de contamination des aliments

(Source: Barro *et al.*, 2007)

Légende : ‑‑‑‑‑‑‑‑‑‑‑▷   Possibilités de relations de contaminations

⟹   Relation de contaminations déjà mis en évidence

III.1. Contaminations d'origine exogène

La contamination d'origine exogène est la présence de contaminants d'origine environnementale sur l'aliment. Les contaminants d'origine exogène proviennent généralement de la flore de l'air, de la flore cutanée ou de la flore de l'eau (Leyral, 1996).

III.1.1. Flore de l'air

L'air est l'une des principales sources de contamination. Il véhicule les microorganismes fixés sur les particules de poussière ou les cellules bactériennes libres (Bello, 2004). Les microorganismes de l'air sont essentiellement des bactéries Gram positif, les spores de moisissures, les microcoques et les staphylocoques (Leyral, 1996). L'air chargé de microorganismes vient au contact de l'aliment lorsqu'il est mal protégé.

III.1.2. Flore cutanée

Les parties externes du corps de l'homme et des animaux sont souvent le site transitoire de plusieurs microorganismes de types commensaux. Au niveau de la peau, les espèces les plus fréquemment rencontrées sont les staphylocoques, les corynébactéries, les levures et les moisissures (Leyral, 1996). Les manipulations non hygiéniques des vendeurs peuvent donc être des sources de contamination par des bactéries pathogènes. Barro *et al.* (2006) ont montré que les mains des opérateurs de la restauration collective et les pièces de monnaie sont porteuses de plusieurs types de germes dont les staphylocoques. Une étude réalisée en France concernant le personnel de cuisine de collectivité a montré que 20 à 30% des individus étaient porteurs chroniques ou occasionnels de *Staphylococcus aureus* qui se retrouve dans l'aliment lors des manipulations (Joffin et Joffin, 1994).

III.1.3. Flore de l'eau

La flore de l'eau est abondante et diversifiée. Les cocci à Gram positif et les bacilles à Gram négatif s'y développent très bien. Plusieurs études ont montré que les eaux de robinets, des puits et des pompes sont souvent contaminées par des coliformes fécaux (Bayanne, 2000; Bello, 2004; Barro *et al.*, 2006). Une étude effectuée à Calcutta a montré que 47% des eaux utilisées dans la préparation des aliments étaient contaminés par les germes fécaux (Chakravarty et Canet, 1996).

III.2. Contamination d'origine endogène

Il s'agit des microorganismes déjà présents dans l'aliment avant sa préparation. Ce sont les agents des zoonoses présents dans les viandes ou les produits carnés et transmissibles à l'homme. Normalement les animaux malades sont écartés par les contrôles vétérinaires mais les animaux porteurs sains existent et peuvent être source de contamination des aliments d'origine animale.

Chapitre 2: Les principales bactéries enteropathogènes transmises par les aliments

Les aliments peuvent être contaminés par plusieurs types de bactéries pathogènes. Les effets indésirables de ces pathogènes chez le consommateur sont caractérisés par des diarrhées, des crampes abdominales et des vomissements, plus rarement des céphalées, des nausées, de la fatigue et de la fièvre (Leyral, 1996). Une étude retroprospective conduite par Barro et Traoré (2001), Barro *et al.* (2005) au Burkina Faso sur les germes impliqués dans les troubles digestifs auprès des services sanitaires a montré que les bactéries les plus fréquentes sont *Escherichia coli* suivies par les Salmonelles, les Shigelles et les Klebsielles.

I. Salmonelles

I.1. Historique

Le bacille a été observé pour la première fois en 1880 par un médecin allemand du nom Eberth (Minor, 1994). L'observation s'est faite sur des sections de rate et de nœuds lymphatiques mésentériques d'un patient mort de typhoïde.

En 1886, Salmon et Smith isolèrent l'actuelle *Salmonella enterica* subsp. *enterica* sérotype *choleraesuis*, autrefois appelée «Bacterium Sui- pestifer » à partir d'un porc atteint de « Hog cholera» (Le Minor, 1994; Yan *et al.*, 2003). Quelques années plus tard, Widal puis Grunbaum, découvrent que le sérum des patients atteints de fièvre typhoïde agglutinait les cultures du bacille de Eberth (bacille de la typhoïde); ce nouveau test est alors appelé le sérodiagnostic de WIDAL. L'organisme isolé est appelé le bacille paratyphoidique par Achard et Bensaude (Grimont *et al.*, 2000).

D'autres bacilles, proches du bacille de la typhoïde et de ceux de la paratyphoïde sont ensuite découverts chez beaucoup d'espèces animales. Dans le passé, les souches de salmonelles isolées de différents hôtes et différentes conditions cliniques étaient considérées comme différentes espèces et les bactériologistes les appelaient aux noms des pathologies qu'elles provoquent ou au nom de l'espèce animale dont le bacille provenait, c'est ainsi qu'on a: *Salmonella* Enteritidis, *Salmonella* Gallinarum, *Salmonella* Abortusovis, *Salmonella* Typhimurium..., puis sont arrivés les noms des lieux où ces germes ont été découverts: *Salmonella* Panama, *Salmonella* Montevideo, *Salmonella* London etc. (Le Minor, 1994).

Dans les aliments, Carlier et Lagrange (2001), rapportèrent que l'observation «princeps» semble avoir été effectuée par le médecin belge Van Ermenghem en 1896. Plutart, le professeur Jean Verge a écrit un mémorable article paru en 1931 sur les toxi-infections alimentaires d'origine carnée et l'inspection bactériologique des viandes.

I.2.Taxonomie et nomenclature

L'étude systématique des antigènes O (de l'allemand Ohne Hauch), flagellaires H (de l'allemand Hauch) et capsulaires K (de l'allemand Kapselle, appelés aussi Vi (pour virulence) ont permis de démontrer qu'il existe 87 facteurs antigéniques O et 96 facteurs antigéniques H (Bouvet, 1995). Les combinaisons des différents déterminants antigéniques entre eux offrent en théorie plus de 20 000 possibilités. L'étude permit aussi de décrire beaucoup d'espèces (sérovars actuels), c'est pourquoi le concept de sérovar égal à une espèce ne pouvait plus tenir car il était devenu impossible de les séparer par les tests biochimiques usuels (Grimont *et al.*, 2000). En 1966, Kauffmann divisait le genre *Salmonella* en quatre sous genres sur la base de réactions biochimiques désignés par des chiffres romains et sans nomenclature formelle (I à IV) et le genre Arizona constituait le sous genre III. Plus tard, Le Minor *et al.* (1970) considéraient que les sous genres de Kauffmann correspondaient bien aux espèces appelées: *Salmonella* Kauffmanii (sous genre I), *Salmonella salamae* (sous genre II), *Salmonella arizonae* (sous genre III) et *Salmonella houtenae* (sous genre IV). La nomenclature bactérienne prit une toute autre considération en 1980, avec la publication des «Approved Lists of Bacterial Names» (Skerman *et al.*, 1980), où les noms de bactéries qui n'apparaissaient pas sur les Approved Lists perdaient leur rang dans la nomenclature. Ces dernières listes avaient compris cinq espèces de salmonelles à savoir: *Salmonella arizonae, Salmonella choleraesuis, Salmonella enteritidis, Salmonella typhi* et *Salmonella typhimurium*.

Dans les trente dernières années, des études basées sur l'hybridation de l'ADN montrèrent que les sous genres I à IV constituaient en fait un groupe d'une seule hybridation d'ADN. Plutard un autre groupe fût identifié avec quelques rares sérovars, c'est le groupe VI ou Bongori (Le Minor *et al.*, 1982, 1986). Le Minor *et al.* (1982) considéraient que tous les sérovars de salmonelles constituaient une seule espèce qui serait *Salmonella choleraesuis* (nom d'espèce du genre *Salmonella*) et qu'elle contenait six sous espèces: *Salmonella choleraesuis subsp. Choleraesuis, Salmonella choleraesuis subsp.salamae, Salmonella choleraesuis subsp.arizonae,*

Salmonella choleraesuis subsp.diarizonae, Salmonella choleraesuis subsp. Houtenae et *Salmonella choleraesuis subsp. Bongori.*

Une nouvelle sous espèce a été rajoutée par la suite, c'est la sous espèce indica (Le Minor *et al.*, 1986). Cette nomenclature qui respecte strictement les règles du code international de nomenclature des bactéries a sérieusement reculé depuis que le nom *de Salmonella choleraesuis* s'est avéré être aussi un nom de sérovar. A partir de là, Le Minor et Popoff (1987) proposèrent le nom de *Salmonella enterica* pour la seule espèce *Salmonella* avec les sous espèces suivantes: *Salmonella enterica subsp.enterica, Salmonella enterica subsp.salamae, Salmonella enterica subsp.arizonae, Salmonella enterica subsp.diarizonae, Salmonella enterica subsp.houtenae, Salmonella enterica subsp.bongori* et *Salmonella enterica subsp. indica.*

La taxonomie se base aussi sur l'espèce génomique, qui est définie maintenant comme un groupe de souches reliées par un taux d'hybridation ADN - ADN supérieur à 70 % avec une instabilité thermique des hybrides inférieure à 5°C (Wayne *et al.*, 1987). Ces hybridations ont montré qu'il n'y avait que deux espèces génomiques dans le genre *Salmonella*: *Salmonella enterica* (espèce habituelle) et *Salmonella bongori* (espèce rare) (Reeves *et al.*, 1989; Yan *et al.*, 2003). *Salmonella enterica* était subdivisée en six sous espèces: *Salmonella enterica subsp.enterica, Salmonella enterica subsp.salamae, Salmonella enterica subsp.arizonae, Salmonella enterica subsp.diarizonae, Salmonella enterica subsp.houtenae* et *Salmonella enterica subsp.indica.*

En 2004, un autre groupe très rare, *Salmonella subterranea*, représentant une autre espèce de *Salmonella* est décrite par Shelobolina *et al.* (2004) et validé par Heyndrickx *et al.* (2005). La nomenclature actuelle qui semble satisfaire la communauté scientifique, en prenant l'exemple de *Salmonella* Typhimurium, s'établit comme suit: Genre: *Salmonella*, espèce: *enterica*, sous espèce: *enterica*, sérotype: Typhimurium ou simplement *Salmonella* Typhimurium (Grimont *et al.*, 2000; Bell et Kyriakides, 2002). Pour des raisons pratiques, les noms attribués aux sérotypes sont conservés pour la sous espèce *enterica*, mais ces noms n'ont pas de statut taxonomique. C'est la raison pour laquelle ils ne sont pas écrits en italique et débutent par une majuscule (Grimont *et al.*, 2000). Les sérovars des autres espèces, autres que *enterica* sont désignés par la formule antigénique de Kauffmann-White, exemple: *Salmonella houtenae*, c'est *Salmonella enterica subsp.houtenae* sérotype 43:z4, z23:- (Bouvet, 1995).

I.3. Caractéristiques biochimiques et antigéniques
I.3.1. Caractéristiques bactériologiques

Les salmonelles sont des bacilles de 0,5 à 1,5 μ x 2,0 à 5,0 μ, à Gram négatif, anaérobies facultatifs, habituellement mobiles grâce à une ciliature péritriche mais des mutants immobiles existent. Ils se développent bien sur les milieux nutritifs ordinaires et donnent en 18 à 20 heures, des colonies de deux à trois millimètres de diamètre à l'exception de certains sérovars donnant toujours des colonies naines (*Abortus ovis* et *Abortus equi*). Ils présentent un G + C % de 50 – 53. Les salmonelles se présentent aussi par des caractéristiques biochimiques communes à l'espèce et se différencient entre elles par d'autres (Tableau I).

Tableau I : Caractéristiques biochimiques des espèces et sous espèces du genre *Salmonella*

Le Genre *Salmonella*	*Salmonella* Enterica						*Salmonella* Bongori
Caractères biochimiques	Subsp. enterica	Subsp. salamae	Subsp. arizonae	Subsp. diarizonae	Subsp. houtenae	Subsp. indica	
O.N.P.G.	-	-	+	+	-	v	+
Gélatinase (36 °C)	-	+	+	+	+	+	-
culture sur milieu KCN	-	-	-	-	+	-	+
Dulcitol fermentation	+	+	-	-	-	V	+
Malonate (utilisation)	-	+	+	+	-	-	-
Sorbitol fermentation	+	+	+	+	+	+	-
Bêta-glucuronidase	v	v	-	+	-	v	-
Alphaglutamyl transférase	v	+	-	+	-	+	+
Lyse par le phage 01.	+	+	-	+	-	+	+

v: variable ou plus tardivement; +: plus de 90 % des souches positives;-: moins de 10 % des souches positives (Source: Grimont *et al*., 2000).

I.3.2. Caractéristiques antigéniques

I.3.2.1. Antigènes somatiques (antigènes O)

Les antigènes O sont sensibles au formol et thermostables comportant deux fonctions support de la toxicité et de la spécificité (Humbert, 1998). Ces antigènes somatiques sont eux aussi divisés en facteurs majeurs et mineurs servant à caractériser les différents groupes antigéniques (exemple : O4 = groupe B, O9 = groupe D…). Les facteurs mineurs ne sont exprimés qu'en présence d'un bactériophage: cette conversion lysogénique n'est fréquente que pour le groupe E. Seuls les antigènes majeurs représentent un intérêt diagnostique. Il existe 87 facteurs pour les antigènes somatiques.

I.3.2.2. Antigènes flagellaires

Les antigènes H sont des polymères de flagelline: protéine de structure des flagelles, qui présente une composition en acides aminés constante pour un type antigénique donné. La flagelline est sous la dépendance de 2 gènes de structure correspondant à la phase 1 et à la phase 2. La majorité des salmonelles sont diphasiques et peuvent donc exprimer alternativement deux spécificités différentes car elles possèdent deux systèmes génétiques. Cependant, un certain nombre de salmonelles sont monophasiques, car à un temps donné, un seul de ces deux gènes s'exprime et les flagelles seront en phase 1 ou 2. De manière aléatoire, toutes les 1000 à 10000 générations, le gène qui n'est pas exprimé s'exprime et l'autre cesse de s'exprimer (Euzeby, 1996).

I.3.2.3. Antigènes d'enveloppe Vi (capsulaires ou antigènes K)

Le seul antigène d'enveloppe reconnu chez les salmonelles est l'antigène Vi (de virulence), qui n'a été identifié que chez trois sérovars: *Salmonella* Typhi, *S.* Paratyphi et *S.* Dublin (quelques souches). Toutes les souches de ces trois sérovars ne possèdent pas cet antigène (Euzeby, 1996; Rycroft, 2000).

I.3.2. 4. Antigènes R et M

Ce sont des antigènes plutôt rares et surtout ne présentant pas d'intérêt pour l'identification des salmonelles. Les bactéries qui possèdent l'antigène R sont avirulent, dérivé de l'antigène O, ses colonies sont rugueuses (forme Rough: R). Il est plus aisément phagocyté et plus sensible aux activités bactéricides cellulaires et sériques. L'antigène M existe essentiellement chez *Salmonella* Paratyphi B et est responsable de l'aspect muqueux des colonies.

I.3.2.5. Fimbriae

Les fimbriae sont des formations en appendices cellulaires, disposés de manière péritriche, d'une longueur variant de 0,2 à 2 µm et dont la largeur est de 7 nm. Ils peuvent être de 100 à 300 unités par cellule bactérienne, alors qu'un fimbriae est constitué d'un millier de sous unités d'un unique polypeptide d'environ 17 Kda, contenant environ 160 acides aminés (Eisenstein, 1996).

I.4. Pathogenicité des salmonelles

L'invasion de l'organisme se fait à travers l'épithélium de l'intestin grêle par la traversée des cellules M des plaques de Peyer conduisant à leur destruction ainsi que le tissu avoisinant. Le génome des salmonelles contient 5 ilots de pathogénicité qui sont absents chez *E. coli*. La plupart des gènes contrôlant l'entrée dans les cellules M sont localisés dans le SPI-1 et codent pour le système de sécrétion de type III, responsable du transport des protéines de la bactérie au cytosquelette des cellules hôtes (Lamont, 2004). Quatre protéines codées par les gènes *SPI*-1, *Sop* E, *Sip*C, et *Stp*P sont injectées dans la cellule hôte et initient le réarrangement du cytosquelette qui entraine le froissement de la membrane cytoplasmique permettant ainsi l'internalisation de la bactérie (Jenkins dans Gillespie et Hawkey, 2006; Bohez *et al.*, 2006). Une fois à l'intérieur du macrophage, *Salmonella* demeure à l'intérieur du phagosome. Les protéines encodées par le gène *SPI*-2, telle la *Spi*C inhibe la fusion des lysosomes et endosomes de la cellule hôte contenant des composants bactéricides avec le phagosome.

D'autres protéines du gène *SPI*-2 interfèrent avec le trafic intracellulaire de la cellule hôte recrutant des métabolites de la synthèse des composants bactéricides pour augmenter les chances de survie de *Salmonella* à l'intérieur du phagosome (Millemann, 1998; Sansonetti et Zychlinsky, 2002).

I.5. Epidémiologie et infection à *Salmonella*
I.5.1. Habitat

La gamme d'hôtes de réservoir des salmonelles est très large. Ces bactéries se retrouvent aussi bien chez les animaux à sang chaud, malades ou porteurs sains (oiseaux, mammifères dont l'homme), que chez les animaux à sang froid (reptiles, poissons et insectes) (Humbert, 1998). Les salmonelles possèdent deux caractéristiques qui expliquent probablement leur très large distribution:

- la diversité des animaux susceptibles de les héberger;
- la capacité de survie des salmonelles dans leur environnement (Bouvet, 1995).

Elles peuvent se retrouver dans le milieu extérieur (sol, eau, aliments pour animaux) ou dans les aliments destinés à l'homme et proviennent en très grande majorité d'une contamination fécale où elles peuvent persister quelque temps et même s'y multiplier suite à des conditions favorables. En effet, elles peuvent survivre de 4 à 9 mois (selon la température: 4 à 20 °C) dans le sol ou en eau d'étang, pendant plus d'un an dans les poussières, jusqu'à 28 mois dans les fientes sèches de volailles, jusqu'à 5 ans dans le duvet de couvoirs et jusqu'à 13 mois sur des carcasses de poulets congelés à - 21°C (Euzeby, 1996).

L'organe principal dans lequel les salmonelles se multiplient activement est constitué par le tube digestif de leurs hôtes potentiels au point qu'ils sont actuellement considérés par certains auteurs comme hôtes normaux du tube digestif et leur seul habitat naturel (sauf *S*. Typhi, *S*. Paratyphi A, B et C, qui sont considérés comme des parasites de l'intestin), et que leur présence ailleurs dans l'environnement ou l'eau ne serait due qu'à des contaminations fécales (Bornet, 2000).

Ainsi tous les animaux sont des porteurs potentiels de salmonelles, qui sont toutes virtuellement dangereuses; leur diffusion dans l'environnement est très importante et on parle de cycle des salmonelles (Bornet, 2000). Chez les poulets, leur lieu de prédilection est constitué par le *caecum*, ce qui explique leur diffusion dans les fientes caecales. Les animaux porteurs sains excrètent de façon intermittente les salmonelles à raison de 10 à 10^7 bactéries par gramme de fécès (Euzeby, 1996; Humbert, 1998).

I.5.2. Spécificité d'hôte

Des critères cliniques et épidémiologiques ont pu être envisagés pour classer les salmonelles. Trois catégories écologiques ont été distinguées selon leurs hôtes préférentiels:

- **les sérovars spécifiques de l'homme:** *Salmonella* Typhi; *S*. Paratyphi A, B, C et *Salmonella* Sendai qui sont responsables respectivement de la fièvre typhoïde et paratyphoïde. Ces maladies font encore des ravages dans les pays en voie de développement où l'hygiène alimentaire est peu ou pas respectée.

- les sérovars spécifiques de certains animaux ou qui peuvent exprimer une certaine pathologie particulière chez certaines espèces animales: exemple, *Salmonella* Dublin chez les bovins (mais aussi chez l'homme), *Salmonella* Abortus ovis, *Salmonella* Abortus equi, *Salmonella* Typhimurium variant Copenhagen chez les pigeons et *Salmonella* Pullorum-Gallinarum chez les volailles (Humbert, 1998).

- les sérovars dits ubiquistes: ils se retrouvent indifféremment chez plusieurs espèces à la fois, c'est le groupe des principaux agents de salmonelloses actuelles pouvant être dangereux pour l'homme et les animaux. Ce sont par exemple *Salmonella* Enteritidis, *Salmonella* Typhimurium, *Salmonella* Infantis et *Salmonella* Saintpaul (Humbert, 1998).

Néanmoins, tous les sérovars sont potentiellement pathogènes pour l'homme et particulièrement responsables de toxi-infections alimentaires ou de portage sain.

I.5.3. Toxi-infections alimentaires (TIA) causées par *Salmonella*

Une toxi-infection alimentaire résulte de l'ingestion d'aliments contaminés par des microorganismes pathogènes ou par leur toxine (Catsaras, 1973). Il ya deux types de TIA qui sont provoquées par l'ingestion d'aliments contaminés par les salmonelles:

- les Toxi-Infections Alimentaires Collectives (TIAC): c'est la survenue dans un espace de temps et d'espace groupé d'au moins deux cas de toxi-infections alimentaires ayant une symptomatologie similaire dont on peut rapporter la cause à une même origine alimentaire. Elles sont à déclaration obligatoire et conduisent à une enquête administrative sous l'égide de la direction de la santé et de la population et la direction des services vétérinaires. Les salmonelles ubiquitaires restent la cause principale des TIAC, déclarées en France (au moins 3 fois sur 4) avec prédominance du sérotype *Salmonella* Enteritidis et une forte progression de *Salmonella* Typhimurium mais aussi *Salmonella* Hadar, *Salmonella* Virchow, *Salmonella* Senftenberg et *Salmonella* Heidelberg. L'augmentation de *Salmonella* Typhimurium est d'autant plus préoccupante qu'il existe une progression de l'antibiorésistance de ce sérotype (Weill, 2006). Les aliments les plus fréquemment en cause sont les oeufs et les ovoproduits, les viandes et les volailles. L'évolution des TIAC est généralement bénigne mais peut être grave aux âges extrêmes de la vie et nécessitent l'hospitalisation, parfois associée à une mortalité non négligeable (Carlier et Lagrange, 2001).

- **les Toxi-Infections Alimentaires sporadiques:** elles sont très fréquentes. Aux Etats Unis d'Amérique, le nombre annuel de ces toxi-infections est de 2.000.000 avec un nombre de morts estimé entre 500 et 2000 (0,05 à 0,10 %). Les aliments incriminés sont les oeufs, les poulets, la viande et d'autres aliments consommés crus (Swerlow et Altekruse, 1998). Les sérotypes dominants sont *Salmonella* Typhimurium et *Salmonella* Enteritidis avec des pourcentages respectifs de 35 et 34 %, alors que *Salmonella* Hadar ne représente que 7%. Il faut signaler que parmi les souches de *Salmonella* Typhimurium isolées partout dans le monde, il existe une variété prédominante caractérisée par son sous type lysotypique: le type DT104 (Definitive Type 104), retrouvé pour plus de 70% des souches de *Salmonella* Typhimurium isolées. Cette souche est caractérisée par sa résistance étendue aux antibiotiques usuels, ce qui explique en partie le facteur de risque associé à la prise d'antibiotiques dans le mois précédant l'épisode de diarrhée infectieuse permettant l'implantation plus facile chez ces enfants ayant une flore digestive fragilisée et incapable d'établir une barrière efficace contre les salmonelles (Carlier et Lagrange, 2001).

II. Escherichia coli
II.1. Caractères généraux
Les *E. coli* sont principalement caractérisées par des critères biochimiques et antigéniques.

II.1.1. Caractères biochimiques
C'est en 1885 que la bactérie *Escherichia coli* a été décrite pour la première fois dans des selles de nourrissons par l'Allemand Theodor Escherich. Toutefois, son nom actuel lui a été donné en 1919 par Castellani et Chambers (Grimont, 1987). Le genre *Escherichia* appartient à la famille des Enterobacteriaceae, qui doit son nom à

leur isolement fréquent du tube digestif et/ou des fèces des mammifères (Greatorex et Thorne, 1994). Les genres constituant cette famille sont des bacilles à Gram négatif, aéro-anaérobies facultatifs qui peuvent reduire les nitrates et qui ne possèdent pas d'oxydase (Le Minor *et al.*, 1990). Le genre *Escherichia* regroupe cinq espèces: *E. blattae, E. coli, E. fergusonii, E. hermanii* et *E. vulneris*. Les membres d'une même espèce présentent habituellement plus de 70% d'homologies génomiques alors qu'entre des espèces différentes, l'homologie est inférieure à 60%. Chaque espèce de *Escherichia* possède des caractéristiques biochimiques spécifiques permettant ainsi de les différencier (Tableau II).

Tableau II: Principaux critères différentiels des espèces du genre *Escherichia* (*E. coli, E. hermanii, E. vulneris, E. fergusonii*)

Carcatéristiques	*E. coli* nonO157:H7	*E. coli* O157:H7	*E. hermanii*	*E. vulneris*	*E. fergusonii*
Indole	+	+	+	-	+
Pigment jaune	-	-	+	(+)	-
LDC	(+)	(+)	-	+	+
ODC	+/-	+/-	+	-	+
β-xylosidase	-	-	-	+	-
β-glucuronidase	(+)	-	-	-	-
Sorbitol	+	-	-	-	-
Malonate	-	-	-	+	-
Adonitol	-	-	-	-	+

+, positif avec une minorité de souche; (+), positif avec la majorité des souches; +/-, positif ou négatif selon les souches; LDC, Lysine Décarboxylase; ODC, Ornithine Décarboxylase

(Source: Grimont, 1987)

II.1.2. Caractères antigéniques

Les antigènes permettent de caractériser les *Escherichia coli* sur le plan phénotypique par utilisation d'antisérum spécifique. On distingue trois types d'antigènes sur *E. coli*: les antigènes O, H et K.

- Les antigènes O ou somatiques comprennent 180 types antigéniques détectables par agglutination;
- Les antigènes H ou flagellaires sont de nature protéique; au nombre de 56, ils sont difficiles à mettre en évidence. Ils ne sont présents que chez les souches mobiles;
- Les antigènes K ou capsulaires sont de nature polysaccharidiques. On distingue actuellement 93 antigènes K de structure polysaccharidique; les souches pathogènes possèdent l'antigène K1.

II.2. *Escherichia coli* et pouvoir pathogène

Escherichia coli est considéré comme faisant partie de la microflore bactérienne du tractus digestif de l'homme ainsi que de celle de nombreux animaux à sang chaud. Elle représente près de 80% de la microflore aérobie (Greatorex et Thorne, 1994). A ce titre *E. coli*, et plus largement les coliformes thermotolérants, sont recherchés dans les aliments comme indicateurs de contamination fécale; leur présence fournit ainsi une indication sur une éventuelle contamination de l'aliment par des bactéries pathogènes d'origine digestive (e.g. *Salmonella typhimurium*, *E. coli* O157:H7). En outre, bien que la majorité des souches de *E. coli* soient commensales banales, certaines d'entre elles sont pathogènes et connues des médecins comme étant à l'origine de pathologies intestinales (Levine, 1987) ou extra-intestinales (Pohl *et al.*, 1989). Les principaux pathotypes intestinaux sont décrits en fonction des facteurs de pathogénicité et des signes cliniques engendrés.

II.2.1. Classification des pathotypes de *Escherichia coli* et leur pathogénicité

Divers chercheurs ont réussi à mettre en évidence pour les souches pathogènes, l'existence de propriétés particulières, dites de virulence, directement ou indirectement reliées à leur pouvoir pathogène. Ces propriétés permettent aux bactéries de coloniser les surfaces muqueuses de l'hôte, de les franchir, de résister aux défenses internes ou de produire un effet toxique sur cet hôte, avec apparition de lésions et de signes cliniques (Cooke, 1985; Sussman *et al.*, 1996). En 1945, un groupe particulier de souches, aujourd'hui appelé *Escherichia coli* entéropathogène (EPEC), fut reconnu comme responsable de cas épidémiques de gastro-entérite chez les jeunes enfants. Depuis, d'autres groupes ont été individualisés et même s'il n'existe pas de classification standardisée des différents pathotypes de *E. coli*, chaque pathotype a reçu un nom basé sur la reconnaissance d'une propriété *in vivo* ou *in vitro*, liée directement ou indirectement aux lésions et à la clinique. Les médecins utilisent une classification basée sur la pathogénie des syndromes diarrhéiques comprenant 6 groupes (figure 2).

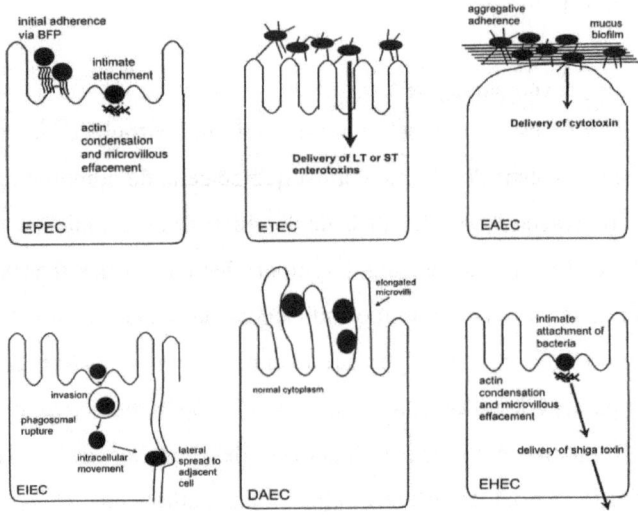

Figure 2: Classification des pathovars de E. coli (●) basée sur leur pathogénicité sur les cellules epithéliales (□) (Source: Nataro et Kaper, 1998).

II.2.1.1. *Escherichia coli* Entérotoxinogènes (ETEC)

Les *E. coli* Entérotoxinogènes (ETEC) sont majoritairement associées à 2 syndromes cliniques importants: les diarrhées du nourrisson dans les pays en voie de développement et la diarrhée du voyageur (ou « turista »). Les ETEC colonisent essentiellement la partie proximale de l'intestin grêle, grâce à leurs « facteurs de colonisation » (CFAx et CSx) qui sont des adhésines fimbriales (Cassels et Wolf, 1995). Elles n'induisent pas d'altération histologique marquée de la muqueuse. Le pouvoir pathogène des ETEC s'explique principalement par la sécrétion des toxines thermostables (ST) et/ou thermolabiles (LT) (Levine, 1987).

II.2.1.2. *Escherichia coli* Entéropathogènes (EPEC)

Les EPEC sont responsables de gastro-entérites infantiles. Elles touchent en particulier les enfants en bas âge (< 3 ans). Lors d'infections, des lésions histopathologiques particulières apparaissent, appelées lésions d'attachement et d'effacement (lésions A/E) (Andrade *et al.*, 1989). Ce phénotype est caractérisé par l'effacement des microvillosités intestinales et par l'adhésion intime entre les bactéries et la membrane cytoplasmique des entérocytes. Plusieurs gènes dont le gène *eae* sont à l'origine de ces lésions (Jerse *et al.*, 1990).

II.2.1.3. *Escherichia coli* Entéroinvasives (EIEC)

Ce pathovar est phylogénétiquement assez proche de *Shigella* spp. (Brenner *et al.*, 1973), et leurs mécanismes de pathogénie sont similaires (invasion de l'épithélium intestinal). Il est responsable de syndromes dysentériques caractérisés par une forte fièvre, des crampes abdominales et des nausées accompagnées d'une diarrhée aqueuse qui évolue rapidement en une dysenterie (des selles contenant du sang et du mucus).

II.2.1.4. *Escherichia coli* Entérohémorragiques (EHEC)

Les EHEC sont à l'origine de troubles plus ou moins sévères allant d'une « simple » diarrhée peu hémorragique à des colites hémorragiques; voire à un Syndrome Hémolytique et Urémique chez l'enfant ou à un Purpura Thrombotique

Thrombocytopénique (PTT) chez l'adulte, pouvant conduire parfois à la mort du patient (Riley *et al.*, 1983). Les EHEC qui correspondent aux souches isolées chez les malades possèdent typiquement au moins un gène *stx* (*stx1* codant la Shiga-toxine 1 [Stx1] ou *stx2* codant la Shiga-toxine 2 [Stx2]) (Riley *et al.*, 1983; Paton et Paton, 1998) ainsi que d'autres facteurs de virulence comme le gène *eae* codant le facteur d'attachement et d'effacement (Levine et Edelman, 1984) ou le gène *ehx* codant l'entérohémolysine (Schmidt *et al.*, 1995). L'ensemble des souches de *E. coli* possédant au moins un gène *stx* représente le groupe des STEC (*E. coli* productrice de Shiga-toxine) ou VTEC (*E. coli* productrice de verotoxine) selon l'ancienne dénomination internationale.

Différentes hypothèses ont été développées sur l'origine des EHEC, sachant que les relations clonales entre les EPEC et les EHEC sont assez complexes, et que de nombreuses souches appartiennent à des sérotypes communs aux deux pathovars. Les EHEC seraient un groupe de clones dérivant des *E. coli* Entéropathogènes (EPEC) (Whittam *et al.*, 1993). Ceci s'expliquerait par le fait que des gènes comme le gène d'attachement et d'effacement (*eae*) auraient été transférés d'une souche pathogène à une souche de *E. coli* commensale qui serait devenue à son tour pathogène.

L'analyse moléculaire et la comparaison de la distribution des gènes spécifiques de virulence ont montré que les EHEC auraient probablement acquis récemment la plupart de leurs facteurs de virulence par transfert horizontal de matériel génétique. L'explication logique de l'émergence des EHEC comme un agent pathogène majeur serait la présence de ces gènes sur des éléments mobiles, comme les gènes *stx1* et *stx2* situés sur des prophages, le gène *eae* (intimine) sur un îlot de pathogénicité *LEE* et le gène *ehx*A et *kat*P (entérohémolysine-EHEC, *Kat*P) sur le plasmide EHEC (pO157), qui auraient été transmis horizontalement en différentes étapes. L'acquisition de gènes *stx* par deux clones EPEC aurait conduit à l'émergence de deux lignées majeures de EHEC: EHEC-1 constituée de souches O157, et EHEC-2 constituée de souches O26 et O111 (Sperandio *et al.*, 1998; Wieler *et al.*, 1997).

II.2.1.5. *Escherichia coli* **Entéroaggrégatives (EAEC)**

Elles sont de plus en plus reconnues comme étant responsables de retards de croissance et de diarrhées persistantes dans les pays en voie de développement ainsi que dans les pays industrialisés. Une adhésion agrégative en " briques empilées", à l'origine de nécroses au pôle apical des villosités avec un oedème inflammatoire et hémorragique de la sous-muqueuse, les caractérise. Elles élaborent une entérotoxine thermostable (EASTI) et une entérotoxine thermolabile (Kaper *et al.*, 2004).

II.2.1.6. *Escherichia coli* **à adhésion diffuse (DAEC)**

Les DAEC sont responsables de diarrhées et d'infections urinaires. Les diarrhées peuvent être aqueuses et contenir du mucus. L'expression d'une adhésine fimbriale (appartenant à la famille Afa/Dr) et d'une protéine de membrane externe confère aux bactéries un phénotype d'adhésion «diffuse» sur les lignées cellulaires en culture de type Hep-2 (Cookson et Nataro, 1996).

II.3. Phylogénie de *Escherichia coli*

Ochman et Selander (1984) ont constitué une collection de 72 souches *E. coli* d'origine diverse (ECOR). La variabilité génétique au sein de la collection a été analysée et les souches ont été classées en 6 groupes phylogénétiques majeurs A, B1, B2, C, D et E. Une étude réalisée sur les souches de la collection ECOR et de la collection DEC (comprenant 15 souches responsables de diarrhées) a révélé une diversité des *E. coli* pathogènes parmi les groupes de la collection ECOR (Boyd et Hartl, 1998). Lors de cette étude, la comparaison des souches portait sur l'analyse des régions spécifiques des îlots de pathogénicité mis en évidence chez les *E. coli* uropathogènes. Il s'est avéré que ces souches ainsi que les autres *E. coli* pathogène extra-intestinales (ExPEC) appartiennent majoritairement aux groupes B2 et D. Les auteurs avaient alors suggéré que les îlots de pathogénicité avaient été acquis par les groupes B2 et D puis par transfert horizontal par certaines souches des groupes A, B1 et E.

Cependant, les souches entéropathogènes sont massivement présentes dans les groupes A et B1, en particulier, les souches des pathovars EHEC, ETEC (entérotoxigéniques) et EIEC (entéroinvasives). Les pathovars EPEC, EAEC (entéroaggragatives) dont les pathologies associées sont plus modérées sont répartis sur l'ensemble des groupes (Escobar-Paramo *et al.*, 2004). Les souches STEC/EHEC sont présentes dans les groupes phylogénétiques A et B1 mais les souches de sérotype O157:H7 sont regroupées dans le groupe E (Girardeau *et al.*, 2005; Escobar-Paramo *et al.*, 2004).

II.4. Recombinaison génétique chez *Escherichia coli*

Escherichia coli est une bactérie qui échange facilement l'information génétique par un transfert horizontal de gène (e.g. par conjugaison, transduction ou transformation) entre d'autres *E. coli* ou d'autres bactéries telles que *Salmonella* ou *Shigella* (EFSA, 2011). Le transfert de gène entre pathovars de *E. coli* se fait essentiellement par la conjugaison. Elle est un transfert d'ADN entre une bactérie donatrice et une bactérie réceptrice, qui nécessite le contact et l'appariement entre les bactéries, et repose sur la présence dans la bactérie donatrice ou mâle, d'un facteur de sexualité ou de fertilité (facteur F qui est un plasmide). Celui-ci permet la synthèse de pili sexuels et donne la polarité au chromosome. Le transfert d'ADN chromosomique qui est à sens unique, orienté, progressif et quelque fois total, a beaucoup de similitudes avec le transfert d'ADN extra-chromosomique. Le transfert d'ADN chromosomique par conjugaison ne se produit qu'entre bactéries d'une même espèce, et surtout chez les bactéries à Gram négatif telles que les entérobactéries (*E. coli*, *Salmonella* spp,...). Le transfert de gènes par conjugaison est un facteur majeur d'évolution du patrimoine génétique bactérien, qui joue un rôle essentiel en bactériologie médicale (surtout la résistance aux antibiotiques). La bactérie impliquée dans la récente épidémie en Allemagne résulte de la recombinaison entre STEC et EAEC (Mellmann *et al.*, 2011; EFSA, 2011).

II.5. Epidémiologie et infection à *Escherichia coli*

Les *E. coli* peuvent être isolées de l'intestin de nombreuses espèces animales. Les animaux peuvent constituer un réservoir et la dissémination dans l'environnement provient essentiellement de contaminations fécales. Peuvent aussi être contaminés l'eau, les sous produits d'activités agro-alimentaires et les aliments pour animaux. Dans le cas des EHEC, les bovins et les ovins, semblent constituer les principaux réservoirs (Hussein, 2007). Le risque de la contamination des denrées alimentaires d'origine animale est fonction de l'importance du portage animal, mais également du respect des procédures d'hygiène appliquées notamment en abattoir et dans les ateliers de transformation. Il existe plusieurs pathovars de *E. coli* que l'on retrouve en pathologie humaine et qui présentent une entéro-virulence. Dans les années 80, les EHEC et plus particulièrement le sérotype O157:H7 sont devenus des pathogènes émergents responsables à la fois de cas sporadiques et de cas groupés de diarrhées souvent sanglantes pouvant évoluer vers des pathologies plus graves comme le syndrome hémolytique et urémique (SHU) et le purpura thrombotique thrombocytopénique (PTT). Différents EHEC autres que *E. coli* O157:H7 sont actuellement rapportés comme producteurs de shiga-toxine. En exemple de *E. coli* non-O157 producteur de shiga-toxine (STEC) nous pouvons inclure O55, O111, O26, O103:H2, O148:H8 et O126 (Bettelheim *et al.*, 2003; Tarr *et al.*, 2005). Les pathovars de *E. coli* sont à l'origine de toxi-infections alimentaires graves principalement transmises par les viandes hachées crues ou mal cuites. Par exemple l'importance de EHEC sur la santé publique est apparue en 1982, à la suite d'une flambée de toxi-infection alimentaire collective aux Etats-Unis d'Amérique. Cette bactérie a provoqué des décès au cours de ces dernières années (au Japon, aux Etats-Unis, au Canada, en Ecosse et en France). En France, depuis 1996, la surveillance des SHU est réalisée chez l'enfant de moins de 15 ans, jusqu'à 2003, 679 cas de SHU ont été notifiés (Brugere et Bailly, 2006). Les *E. coli* sont fréquemment isolées des aliments d'origine animale tels que la viande hachée crue ou mal cuite, les volailles insuffisamment cuites, le lait cru ainsi que dans l'eau non chlorée et les jus non pasteurisés d'une façon générale dans les aliments préparés sous des mauvaises conditions d'hygiène (Barro *et al.*, 2002; Blanco *et al.*, 2004).

II.5.1. Présentation de certaines épidémies à STEC liées aux aliments carnés

En Novembre 1994, 17 personnes sont tombées malades suite à l'ingestion de salami fermenté et séché, prétranché, commercialisé à Washington et en Californie. Suite à cette épidémie, *E. coli* O157:H7 a été isolée dans du salami encore emballé directement collecté dans les supermarchés ayant approvisionné les 2 Etats (Tilden *et al.*, 1996). Il y a eu une identité totale démontrée par électrophorèse en champ pulsé entre les souches isolées des malades et celles isolées du salami. En 2002, une épidémie d'envergure associée à la saucisse fermentée contaminée par *E. coli* O157:H7 a été identifiée dans la région sud de la Suède: 39 cas d'infection ont été répertoriés et parmi eux, 12 cas de très jeunes enfants ayant développé un syndrome hémolytique et urémique (Sartz *et al.*, 2008). Dans ce cas également, il y a eu la preuve de la relation clonale entre les souches isolées de la saucisse fermentée et celles isolées des selles des malades. La viande de bœuf, matière première de fabrication de ces saucisses a été suspectée comme étant la source de l'infection. Ainsi, pour assurer la sécurité sanitaire de ces produits carnés fermentés, différents pays dont les États-Unis, l'Australie et le Canada, ont mis en place des mesures renforcées de contrôle intégrant notamment comme support d'aide à la gestion, la microbiologie prévisionnelle qui permet de déterminer la survie et la croissance de différents pathogènes dont *E. coli* O157:H7 dans ces produits fermentés carnés (Anon, 1999).

Même si le sérotype O157:H7 est un sérotype majeur dans l'émergence d'épidémies d'envergure, il n'en est pas moins vrai que d'autres sérogroupes, non O157 ont été cités dans des épidémies liées à des produits carnés fermentés. Ainsi, une épidémie ayant mis en cause des saucisses fermentées séchées contaminées par *E. coli* O111 a été identifiée à Adélaïde, région sud de l'Australie entre janvier et février 1995 avec 21 syndromes hémolytiques et urémiques (SHU) identifiés. La viande, matière première utilisée pour fabriquer ces saucisses était issue de moutons, porcs et agneaux. Au Danemark en 2007, une épidémie liée à des saucisses de bœuf contaminées par *E. coli* O26:H11 a été décrite. Cette épidémie a fait 20 malades dont 3 SHU, la majorité des malades étant composée d'enfants (Ethelberg *et al.*, 2007). Des études de traçabilité ont permis de retrouver le lot de viande de bœuf contaminé, de rappeler et de détruire les saucisses contaminées.

En 2008, mais cette fois-ci en Norvège, ce sont des saucisses de mouton séchées et fermentées contaminées par *E. coli* O103:H25 qui ont été à l'origine de 6 cas de SHU tous affectant de très jeunes enfants (Schimmer *et al.*, 2008). Pour cette épidémie également, il a pu être démontré un lien de clonalité entre les souches excrétées à l'échelon d'une de ces 4 fermes et les souches isolées à la fois dans les saucisses fermentées de mouton et les selles des jeunes enfants victimes d'insuffisance rénale.

Dans les pays en développement, plusieurs cas d'épidémies sont signalés par la presse chaque année suite à la consommation d'aliments contaminés. Malheureusement, les bactéries responsables ne sont pas recherchées dans les aliments en cause par faute d'investigation épidémiologique.

II.5.2. Principales étapes du processus infectieux des STEC

Après ingestion, les STEC doivent résister à l'acidité de l'estomac. Une étape de colonisation du tube digestif est probablement nécessaire: la plupart des souches STEC (en particulier celles de sérotype O157:H7) sont capables de produire des lésions d'attachement/effacement; pour les autres, les mécanismes de colonisation sont encore mal connus. Les toxines produites par les bactéries doivent ensuite traverser l'épithélium intestinal, avant de rejoindre le système circulatoire et atteindre les récepteurs spécifiques (Gb3) localisés à la surface des cellules endothéliales, principalement au niveau intestinal, rénal et cérébral. Les toxines Stx entraînent la mort des cellules cibles par arrêt de synthèses protéiques (figure 3). Un rôle des bactéries et/ou des toxines sur l'activation du système immunitaire est également suspecté (Heyderman et *al.*, 2001).

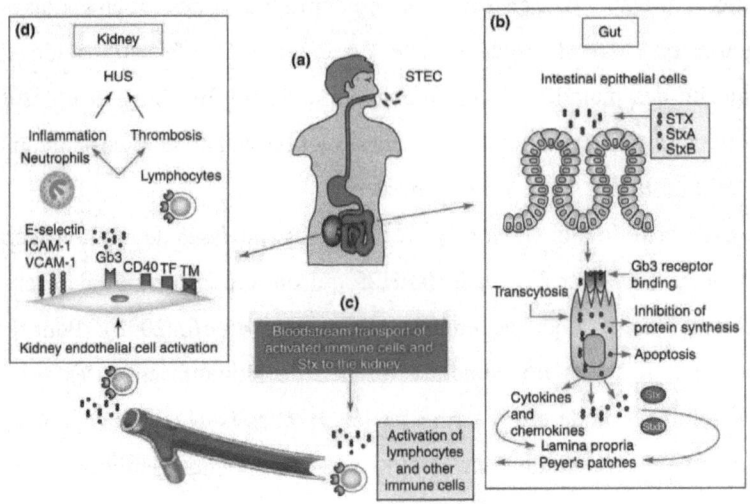

Figure 3: Principales étapes du processus infectieux des STEC (Heyderman *et al.*, 2001)

(a): Passage de STEC dans la bouche; (b): processus infectieux des STEC dans les intestins

(c): processus infectieux des STEC dans le sang; (d): processus infectieux des STEC dans les reins

III. Autres espèces bactériennes transmissibles par les aliments
III.1. Campylobacter spp

Le genre *Campylobacter* tout comme le genre *Salmonella* est la principale cause de toxi-infection alimentaire à travers le monde (Mead *et al.*, 1999). Ce genre renferme des espèces pathogènes qui sont *C. jejuni* et *C. coli* qui causent plus de 95 % des campylobactérioses. La consommation de viande de volaille insuffisamment cuite est la principale cause de campylobactériose sporadique (Pearson *et al.*, 1993). Le genre *Campylobacter* est constitué de bactéries en forme de fins bacilles à Gram négatifs incurvés en spirale, non sporulés, parfois en forme de S, d'une taille de 0,2 à

0,5 µm de large et de 0,5 à 5 µm de long (Leyral, 1996). Les bacilles sont mobiles grâce à un flagelle situé à une ou aux deux extrémités de la cellule et ont un mouvement typique de tire-bouchon. *Campylobacter* a un métabolisme de type respiratoire et est micro aérophile, c'est-à-dire qu'il requière une concentration en oxygène entre 3 et 15 % (Ghafi et Daube, 2007). Certaines souches peuvent occasionnellement se multiplier dans des conditions d'aérobiose ou d'anaérobiose. Ils sont incapables d'oxyder ou de fermenter les sucres et sont positifs au test de l'oxydase (Leyral, 1996). Toutes les espèces de *Campylobacter* se multiplient à 37°C, mais les *Campylobacter* thermophiles (*C. jejuni, C. coli* et *C. lari*) ont une meilleure croissance à 42°C et ne se multiplient pas à une température inférieure à 25°C. Ces bacilles sont plus sensibles aux conditions défavorables, telles que la dessiccation, la chaleur, l'acidité, les désinfectants ou l'irradiation que la plupart d'autres bactéries pathogènes intestinales (Hu et Kopecko, 2003). Le réservoir de ces bactéries est le tractus intestinal des animaux domestiques (porcs, bovins, ovins, volailles) et sauvages particulièrement les oiseaux (Butzlert, 2004). La transmission a lieu généralement par la consommation d'aliments (viande de volaille insuffisamment cuite qui est la principale source de contamination de l'homme), d'eau, des contacts directs ou la manipulation d'animaux infectés (animaux de boucherie et de compagnie) (Hu et Kopecko, 2003). La campylobactériose se manifeste par la fièvre, la diarrhée et de fortes douleurs abdominales (Jorgensen *et al.*, 2002).

III.2. *Yersinia* spp

Le genre *Yersinia* comprend 11 espèces appartenant à la famille des *Enterobacteriaceae*. Il s'agit de bacilles Gram négatifs, non sporulés, anaérobies facultatifs qui fermentent le glucose. Plus petites que la plupart des autres entérobactéries, elles apparaissent souvent comme des coccobacilles lorsqu'elles se multiplient à 37°C. Ce genre comprend 4 espèces pathogènes bien caractérisées: *Yersinia pestis* responsable des pestes bubonique et pulmonaire, *Y. pseudotuberculosis* pathogène des rongeurs et occasionnellement de l'homme, *Y. ruckeri* provoquant des maladies chez les poissons d'eau douce, et *Y. enterocolitica*, un pathogène intestinal. *Y. pseudotuberculosis* et *Y. enterocolitica* sont les 2 agents pathogènes d'origine alimentaire (Ghafi et Daube, 2007). Elles atteignent le tractus gastro-intestinal de l'homme et provoquent des entérites, entérocolites, lymphadénites et rarement des infections extra-intestinales telles que des arthrites. *Y. enterocolitica* est également présente dans l'intestin d'animaux sains tels que les porcs, bovins, chiens et chats (Krauss *et al.*, 2003). L'espèce *Y. enterocolitica* est divisée en plusieurs sous-groupes suivant leur activité biochimique (biotype) et les antigènes O lipopolysaccharides (sérotype) qu'ils portent. L'infection du tube digestif a pour origine l'adhésion et l'invasion des cellules de la lumière intestinale.

III.3. *Shigella* spp

Elle est la cinquième cause de toxi-infections en France. Le genre *Shigella* est constitué de quatre espèces qui se différencient par leur antigène somatique "O". Ce sont *Shigella dysenteriae*, *Shigella boydii*, *Shigella flexneri* et *Shigella sonnei* (Leyral, 1996). Les Shigelles sont responsables d'une dysenterie aiguë qui se caractérise par une fièvre élevée, l'émission fréquente de selles glairo-sanglantes purulentes ou hémorragiques, de vomissements et des douleurs abdominales violentes. La sévérité de la maladie et le taux de décès dépendent de l'hôte et du serovar; les formes graves sont associées à *S. dysenteriae* et *S. flexneri* (Cheftel *et al.*, 1997). Dans les pays en développement *Shigella dysenteriae* est la plus fréquente, la contamination se fait par un contact de l'aliment avec les matières fécales. Les aliments incriminés sont: les viandes, les produits laitiers et l'eau (Joffin et Joffin, 1994).

III.4. Staphylococcus aureus

Les infections à *Staphylococcus aureus* constituent la deuxième cause de toxi-infections à travers le monde. De nombreuses souches sont productrices d'entérotoxines qui sont partiellement thermostables, et sont produites dans les aliments lors de la croissance du germe (Leyral, 1996). La symptomatologie est brutale; après une période courte d'incubation apparaissent de façon brusque et violente des céphalées, des douleurs abdominales, des nausées et des vomissements violents. Les aliments en cause sont des produits contaminés après la cuisson: viande, poisson, divers plats cuisinés (Leyral, 1996).

IV. Situation des entéropathogènes isolés des aliments au Burkina Faso

Des études antérieures ont rapportés la présence de staphyloccoques et d'entérobactéries dans les aliments au Burkina Faso, avec une prédominance de *Shigella, Salmonella* et *E. coli* (Barro *et al.*, 2002; Barro *et al.*, 2005). Ces études ont signalé plusieurs cas de toxi-infections alimentaires souvent mortels dont la presse avait fait l'écho. Ces infections dues au manque d'hygiène sont également ressenties au niveau des services de santé qui soulignent que les diarrhées constituent la troisième cause de consultation en milieu hospitalier (Ministère de la santé, 2007).

Cependant, aucune étude ne s'est interessée sur l'identité sérotypique de *Salmonella* ou de l'identité des pathovars de *E. coli* au niveau des aliments qui sont reconnus comme principale cause de toxi-infections alimentaires et des maladies diarrhéiques au Burkina Faso (Barro *et al.*, 2007). Il est donc nécessaire de disposer de données scientitifiques sur la prévalence de ces pathogènes dans les aliments pour la mise en place d'un réseau national de sécurité sanitaire des aliments.

Chapitre 3: Antibiorésistance des bactéries entéropathogènes

I. Caractéristiques des antibiotiques

Les antibiotiques sont des substances antimicrobiennes qui empêchent la multiplication des bactéries (bactériostatique) ou qui les tuent (bactéricide). En bactériologie médicale, on définit les antibiotiques comme étant des composés chimiques, élaborés par un microorganisme ou produit par synthèse ou par hémi-synthèse et dont l'activité spécifique se manifeste à faible dose sur les microorganismes (Lagrance et Reinert, 1987).

Les antibiotiques peuvent être classés selon leur origine, structure chimique, leur mode/ mécanisme/ site d'action (Figure 4) et leur spectre d'activité (large, étroit).

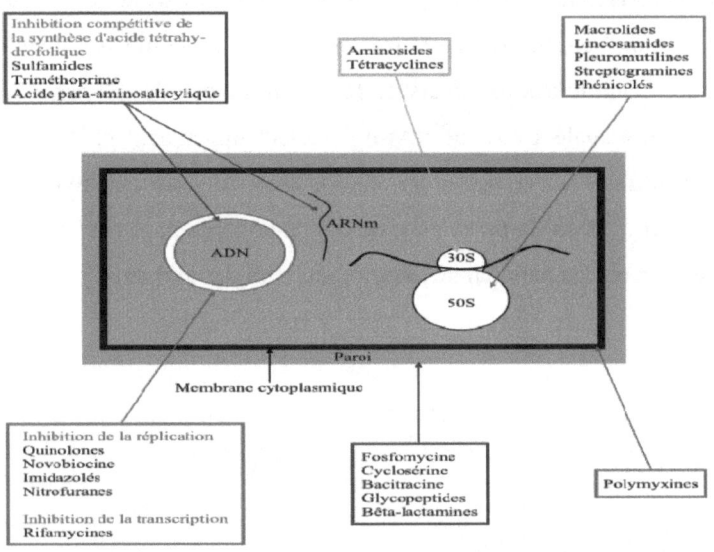

Figure 4: Sites d'action des antibiotiques (Source: http://pedagogie.ac-montpollier.fr)

Une classification des antibiotiques basée sur le site d'action dans la bactérie ou sur le processus physiologique est d'un intérêt bactériologique (Tableau III). A l'intérieur de chacun de ces groupes, la classification par famille est fondée sur la structure chimique des différentes molécules. Et à chaque famille correspond un mécanisme moléculaire spécifique (François *et al.*, 2003).

Tableau III : Classification des antibiotiques selon leurs sites d'action

Sites d'action des antibiotiques	Familles d'antibiotiques	Antibiotiques
Antibiotiques agissant sur les parois	β-lactamines	Pénicillines (Pénicilline G) Céphalosporines (Cefalotine) Monobactams (Aztréonam)
	Fosfomycines	
	Glycopeptides	Fosfocine
		Vancomycine
Antibiotiques agissant sur la membrane	Polymixine	Polymixine B ; Polymixine E
	Gramicidines et Tyrocidines	Bacitracine ; Tyrothricine
Antibiotiques agissant sur le Ribosome	Aminosides	Gentamycine
	Macrolides- Lincosamides- Synergistines	Erythromycine, Lincomycine Virginiamycine
	Phénicols	Chloramphénicol
	Tétracyclines	Tétracycline
	Acides Fusidiques	Fucidine
	Oxazolidinones	Linézolide

Antibiotiques agissant sur l'ARN-polymérase	Rifampicines	Rifamycine
Antibiotiques agissant sur l'ADN	Quinolones	Acide nalidixique
	Produits Nitrés	Nitroxoline
Antibiotiques agissant sur la synthèse de l'acide folique	Sulfamides	Sulfadiazine
	Triméthoprime	Triméthoprime

II. Utilisation des antibiotiques en médecine vétérinaire

Les antibiotiques sont utilisés en médecine vétérinaire de quatre façons avec des objectifs différents (Schwarz *et al.*, 2001):

* **usage thérapeutique** : l'objectif majeur est d'obtenir la guérison des animaux cliniquement malades et d'éviter la mortalité.

* **usage métaphylactique** : lorsqu'une infection collective et contagieuse se déclare chez quelques animaux dans des élevages avec de grands effectifs, l'ensemble du groupe des animaux est traité. Cet usage permet de traiter les animaux soumis à la pression infectieuse alors qu'ils sont encore en incubation ou lorsque les manifestations cliniques sont très discrètes, ce qui permet un traitement collectif par voie orale.

* **usage prophylactique** : les antibiotiques peuvent parfois être administrés à des périodes critiques de la vie des animaux soumis à une pression de contamination régulière et bien connue. Dans la filière porcine, des antibiotiques peuvent être donnés à titre préventif au moment du sevrage qui constitue un stress important pour les animaux.

* **facteurs de croissance** : les antibiotiques peuvent être utilisés dans l'aliment à titre d'additifs en vue d'améliorer la croissance et les performances des animaux, sans que les mécanismes à l'origine de l'amélioration de ces performances aient été clairement élucidés (Dibner et Richards, 2005). Cet usage a fait l'objet de nombreuses critiques et il est interdit au sein de l'Union Européenne depuis 2006 (Rossolini *et al.*, 2008).

III. Impact de l'antibiothérapie sur la flore digestive des hôtes réservoirs

Dans le cas d'une antibiothérapie, une fraction de la dose administrée peut être excrétée sous forme active par voie biliaire ou sécrétée par la muqueuse intestinale, à laquelle s'ajoute en cas d'administration orale la fraction non absorbée. Les quantités d'antibiotiques qui sont alors présentes dans le tube digestif peuvent altérer l'équilibre microbien de la flore intestinale. Les effets observés chez l'hôte peuvent être alors (Edlund et Nord, 2000):

- l'élimination des bactéries sensibles aux concentrations actives de l'antibiotique;

- la sélection et la prolifération de bactéries résistantes au sein de la flore endogène.

L'antibiothérapie modifie donc la stabilité de l'écosystème bactérien et altère la composition de la flore responsable de la résistance à la colonisation. Plusieurs facteurs interviennent sur l'intensité des modifications de l'écosystème intestinal au cours d'un traitement antibiotique, notamment la concentration active qui atteint la lumière intestinale, l'activité intrinsèque de la molécule sur les bactéries qui composent l'écosystème, de sa fixation à des composants du bol alimentaire (protéines, cellulose) et son inactivation éventuelle dans le contenu intestinal (Sullivan *et al.*, 2001).

III.1. Emergence de microorganismes pathogènes dans la flore digestive

La disparition de l'effet de barrière peut entraîner l'émergence de bactéries pathogènes. Chez l'homme, les colites pseudomembraneuses, liées au développement de *Clostridium difficile* producteur de toxine, et la pullulation de levures saprophytes telles que *Candida albicans* sont l'illustration de l'effet des antibiotiques sur la flore intestinale (Sullivan *et al.*, 2001).

III.2. Emergence de bactéries résistantes dans la flore digestive chez l'animal et conséquences en termes de santé publique

III.2.1. Emergence de bactéries zoonotiques résistantes

Chez l'animal, le traitement antibiotique peut permettre l'émergence ou la sélection de bactéries zoonotiques résistantes dans le tube digestif des animaux. Ces bactéries peuvent être transmises à l'homme par les aliments qu'il consomme et être responsables de toxi-infections alimentaires. Ces bactéries peuvent être également transmises par contact avec les animaux ou indirectement par d'autres voies. Les bactéries du genre *Campylobacter* et *Salmonella* sont les principales bactéries responsables de gastroentérites d'origine infectieuse et la viande de porc fait partie des sources de contamination de l'homme. Le fait que ces bactéries zoonotiques soient résistantes aux antibiotiques peut avoir plusieurs conséquences (Molbak, 2005):

- En général, les diarrhées bactériennes zoonotiques de gravité faible ou modérée ne sont habituellement pas traitées avec des antibiotiques. Les patients âgés, les patients bactériémiques ou les malades à risque de complication peuvent cependant être traités avec des antibiotiques. Si les bactéries zoonotiques sont résistantes, on peut assister à un échec thérapeutique.

- Les bactéries zoonotiques résistantes seraient à l'origine d'infections plus invasives et d'une augmentation de la mortalité des patients.

- Si des bactéries zoonotiques résistantes contaminent un individu soumis préalablement à une antibiothérapie, elles auront un avantage sélectif pour coloniser le tube digestif de cet individu.

En termes de santé publique, le premier enjeu de l'utilisation des antibiotiques chez les animaux est donc de limiter leur impact sur l'émergence et la sélection de bactéries zoonotiques résistantes dans le tube digestif des animaux car la diffusion de bactéries zoonotiques résistantes de l'animal à l'homme est possible, et de nombreux arguments attestent de sa réalité (Ungemach *et al.*, 2006).

III.2.2. Emergence de bactéries commensales résistantes

Le traitement antibiotique peut aussi permettre l'émergence ou la sélection de bactéries commensales résistantes dans le tube digestif. La dissémination des déterminants génétiques de la résistance portés par une bactérie commensale peut être considérable puisque ces gènes peuvent être transférés verticalement (à la descendance) ou horizontalement (aux autres lignées bactériennes) s'ils sont situés sur des éléments génétiques mobiles. Ces échanges peuvent survenir entre souches résidentes de la flore intestinale provenant d'espèces identiques, différentes voire éloignées (Vedantam et Hecht, 2003). Le tube digestif des animaux devient un réservoir de gènes de résistance à l'antibiotique administré. Les bactéries commensales sont transmises à l'homme par les mêmes voies que les bactéries zoonotiques. Les bactéries commensales d'origine animale peuvent donc potentiellement se retrouver dans le tube digestif de l'homme et même si elles n'y sont que de façon transitoire, elles peuvent alors éventuellement transférer des gènes aux bactéries commensales humaines. L'écosystème intestinal humain se trouverait donc enrichi de déterminants génétiques de résistance d'origine animale, potentiellement transmissibles à des bactéries pathogènes (Salyers *et al.*, 2004).

Les bactéries isolées chez les animaux et chez l'homme partagent les mêmes mécanismes de résistance, ceci est un argument en faveur de l'absence d'étanchéité entre les populations bactériennes d'origine humaine et animale. Les flux de gènes de résistance peuvent à priori s'effectuer dans les deux sens, et l'évaluation des flux de gènes entre les populations bactériennes humaines et animales nécessite de développer des marqueurs qui permettent de reconnaître l'écosystème d'origine de la bactérie ou du mécanisme de résistance. Ces marqueurs sont particulièrement difficiles à définir car les populations bactériennes d'origine humaine et animale peuvent être relativement proches. En effet, une étude récente vient de montrer qu'en Ouganda les populations humaines rurales et le cheptel partageaient des souches de *E. coli* identiques sur le plan génétique (Rwego *et al.*, 2008).

IV. Mécanismes de résistance des bactéries aux antibiotiques

Une bactérie est dite résistante lorsqu'elle est capable de se développer en présence d'un taux d'antibiotique plus élevé que le taux habituel. La résistance peut être constitutive du germe ou acquise par lui au cours de son développement (Courvalin et Phillipon, 1989). Les bactéries peuvent devenir résistantes à l'action des antibiotiques par quatre mécanismes principaux:

- diminution de la perméabilité membranaire;
- altération de la cible moléculaire, après mutation ou modification enzymatique;
- excrétion accrue de l'antibiotique qui entraîne une concentration insuffisante au niveau de la cible intracellulaire;
- inactivation enzymatique de l'antibiotique, qui peut être hydrolysé (pénicillinase, céphalosporinase pour les β-lactamines) ou modifié dans sa structure chimique (acétylase, adénylase, phosphorylase pour les aminosides).

DEUXIEME PARTIE

MATERIEL ET METHODES

Chapitre 1: Sites d'études et échantillonage

I. Caractéristique de la zone d'étude

I.1. Zone et période d'étude

L'étude a été menée à Ouagadougou (Burkina Faso) en collaboration avec Helsinki (Finlande). Le Burkina Faso est un pays enclavé avec 16.241.811 d'habitants et Ouagadougou, sa capitale administrative, concentre près de 1.200.000 d'habitants soit 40% de la population urbaine du pays (INSD, 2006). La Finlande est un pays nordique avec une population de 5.363.624 d'habitants dont Helsinki la capitale compte 1.322.757 d'habitants (Banque mondiale, 2010; http://donnees.Banquemondiale .org /pays/finlande).

La collecte des échantillons et les analyses préliminaires ont été conduites à Ouagadougou, les analyses plus poussées en Finlande.

Figure 5A: Les différentes étapes de l'étude

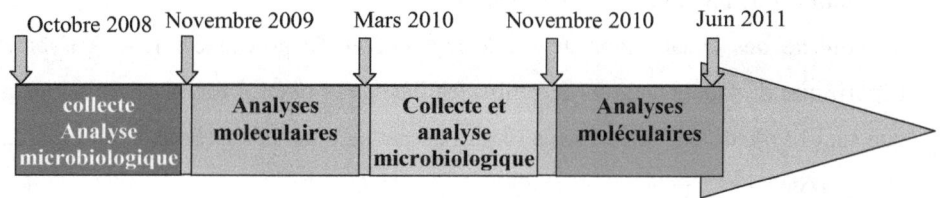

Figure 5B: Localisation géographique du Burkina Faso et de la Finlande

I.2. Sites de collecte et matériel d'étude

La collecte des échantillons a été faite d'abord au niveau des marchés locaux de la ville de Ouagadougou qui sont:

- Le marché populaire de Bendogo sur la route de Fada N'Gourma à la sortie du secteur 27. Sur ce site, seulement les viandes de bœuf et de monton ont été concernées.

- Le marché de Dassasgho au secteur 28, en face de la gare de l'Est; un site d'abattage de poulet, de vente de carcasses crues et grillées a été choisi pour l'étude.

- Le marché de Zogona situé au secteur 13 en bordure de l'avenue Babanguida, un site a concerné la viande de bœuf et boyaux et un autre des carcasses de poulet.

- Le marché de Koulouba au secteur 4 sur la route de l'aéroport, les viandes de mouton et de bœuf ont été prélevées.

La collecte des échantillons de fèces de bœuf et de porc a été faite à l'abattoir frigorifique de Ouagadougou et, enfin, les hérissons ont été reçus de trois localités que sont Mogtédo et Tuiré dans la région du plateau centrale et Fada N'gourma dans la région de l'Est.

II. Enquêtes

Une enquête a été menée sur les conditions de transport des carcasses vers les boucheries. Les conditions d'exposition, de découpe des différentes carcasses et de leur vente ont été scrupuleusement suivies. Une attention particulière a été accordée à l'environnement de vente et aux pratiques hygiéniques des vendeurs (disposition, protection, manipulation et conservation des viandes).

III. Echantillonage

L'échantillonnage a consisté à acheter des portions de viande en tas d'environ 200g. Au total, 120 échantillons constitués de viande de bœuf (n= 45), de mouton (n= 30) et de boyaux de bœuf (n= 45) ont été prélevés dans les conditions naturelles de vente. Puis 130 carcasses fraîches et entières de poulet ont été achetées directement avec les vendeurs de poulet ciblés. L'échantillonage du matériel fécal a été fait sur du bœuf (n= 304), du poulet (n= 350) et du porc (n= 50). Environ 100 à 200g de fèces ont été prélevés au cours de l'éviscération de ces animaux à l'aide de spatule stérile.

Les échantillons ont été à chaque fois placés dans des sachets en plastiques stériles, conditionnés dans une glacière munie de glace et acheminés au laboratoire puis immédiatement analysés dans les 2 h qui suivent chaque prélèvement. Les hérissons (n= 25) ont été capturés vivants et acheminés dans le laboratoire du CRSBAN où ils ont été abattus immédiatement après réception et 1 g des fecès de chaque hérisson a été prélevé pour les analyses microbiolgiques.

Chapitre 2: Analyses bactériologiques et moléculaires

I. Analyses bactériologiques

Les isolements, la caractérisation biochimique et la conservation des souches ont été réalisés au laboratoire de biologie moléculaire et d'épidémiologie de bactéries et virus transmis par les aliments du CRSBAN à l'Université de Ouagadougou.

I.1. Préparation des échantillons

Les préparations des échantillons ont été spécifiques pour chaque type d'échantillon et pour chaque germe recherché. Dans tous les cas, des fractions allant de 1 à 25 g d'échantillon ont été prélevées à l'aide de scalpel ou de spatule stériles et mis dans les sachets Stomacher contenant de l'eau peptonée tamponnée. La préparation a été soumise à l'action du Stomacher (Stomacher 400) pendant 3 - 5 minutes pour libérer les microorganismes.

I.2. Isolement et identification des différents pathogènes recherchés

Ils ont été effectués selon les méthodes standards d'analyses microbiologiques.

I.2.1. *Salmonella* spp

La recherche de *Salmonella* a nécessité quatre étapes successives ci-dessous citées.

-Le pré-enrichissement: 1 à 25 g de viande ou de fèces de chaque échantillon ont été prélevés aseptiquement à l'aide d'un scalpel ou de spatule stérile et placés dans 9 ou 225 ml d'eau peptonée (Difco, BD, Sparks, MD) stérile puis incubés à 37°C pendant 24 h.

- L'enrichissement: 10 µl d'eau peptonée pré-enrichie pour chaque échantillon ont été prélevés à l'aide d'une micropipette et ajoutés dans un tube contenant 10 ml de bouillon Rappaport-Vassiliadis (Oxoid, Basingstoke, Angleterre), puis incubé à 37°C pendant 24 h; ou 2 ml d'eau peptonée pré-enrichie dans du bouillon sélénite cystine (Emapol) puis incubé à 37°C pendant 24 h.

- **L'isolement:** 10 µl du bouillon Rappaport-Vassiliadis (Oxoid, Basingtoke, England) ou du bouillon sélénite cystine (Emapol) enrichi ont été prélevés à l'aide d'une micropipette et ensemencés sur la gélose XLD (Xylose-Lysine-Deoxycholate) (Oxoid, Basingstoke, Angleterre) ou la gélose Hektoen (Difco, BD, Sparks, MD) par la méthode de stries à l'aide d'une pipette Pasteur stérile. Les boites ont été incubées à 37°C pendant 24 h. Après 24 h d'incubation, les colonies suspectes ont été ré-isolées sur XLD ou Hektoen pour la caractérisation biochimique.

- **Identification:** les colonies suspectes qui sont rouges à centre noir sur XLD et bleues ou bleu - vertes et avec ou sans centre noir sur Hektoen ont été caractérisées biochimiquent selon la méthode de Poelma *et al.* (1984). Cette méthode consiste à faire des tests biochimiques suivantes: coloration de Gram, uréase, production d'indole, fermentation du glucose, du lactose et du mannitol, production d'H_2S et de gaz, recherche de mobilité.

La galerie API 20 E a été utilisée pour la confirmation des souches. Les souches ainsi confirmées ont été conservées à -20°C pour être analysées plutard afin de déterminer le sérotype et la suceptibilité aux antibiotiques.

I.2.2. Escherichia coli

Pour la recherche des *Escherichia coli*, 25 g de viande ou de fèces ont été prélevés aseptiquement et mis dans un sachet stomacher contenant 225 ml d'eau peptonée taponnée; puis soumis à l'agitation du Stomacher pendant 2 à 3 minutes. Après homogénéisation, 0,1 ml de la préparation obtenue a été prélevé à l'aide d'une seringue stérile de 1 ml et ensemencé sur la gélose Mueller Hinton (MH) (Liofilchem, Italie) ou la gélose Sorbitol MacCkonkey (SMAC) (Oxoid, Basingstoke, Angleterre). La méthode utilisée est celle du Standard Plate Count qui consistait à une répartition homogène de l'aliquote à l'aide d'une pipette Pasteur recourbée. Les boites ensemencées ont été incubées à 44°C pendant 24 h. Après 24 h d'incubation, la masse bactérienne sur la gélose MH ou SMAC pour chaque échantillon, a été raclée et conservée dans des tubes contenant 1 ml de la gélose trypticase soja à -30°C pour les analyses moléculaires.

I.3. Conservation et envoi des souches pour les analyses moléculaires

Tous les isolats obtenus à partir des différents échantillons ont été conservés dans des cryotubes contenant du bouillon cœur cervelle (BCC) à 15% de glycérol préalablement stérilisés. Les cryotubes ont été ensuite placés dans des cryoboites et conservés à - 30°C pour des analyses ultérieures. Pour l'étape d'envoi, les souches ont été repiquées sur la gélose MH (Liofilchem, Italie) et conservées dans des tubes contenant la gélose tryticase soja, puis emballées selon les normes de l'OMS et envoyées par DHL/Ouagadougou pour la Finlande.

II. Caractérisation phénotypique et moléculaire des souches

Cette partie de notre étude à été entièrement réalisée dans le laboratoire de bactériologie de l'Institut National de santé publique et de bien être de Helsinki, Finlande (THL).

II.1. Caractérisation moléculaire des souches de *E. coli*

En Finlande, les souches ont été ré-isolées sur la gélose CLED (Oxoid, Basingstoke, Angleterre) ou la gélose SMAC (Oxoid, Basingstoke, Angleterre) et incubées à 37°C pendant 24 h.

II.1.2. Réaction de polymérisation en chaine (PCR)

Le principe de la PCR est basé sur l'amplification de manière importante, cyclique et spécifique d'un fragment d'ADN, grâce à une ADN polymérase (*Taq* polymérase: *Thermus aquaticus* polymérase) thermostable, en présence d'amorces spécifiques et de nucléotides. Une PCR multiplex a été utilisée au cours de notre étude appelée 16-plex PCR, développée par Antikainen *et al.* (2009).

II.1.2.1. Détection des pathogroupes de *E. coli* par PCR multiplex (16-plex PCR)

La 16-plex PCR est une méthode qui permet de détecter simultanément la présence de 16 gènes (*uidA, pic, bfp, invE, hlyA, elt, ent, escV, eaeA, ipaH, aggR, stx1, stx2, estIa, estIb* et *a*st) appartenant aux 5 principaux pathogroupes de *E. coli* dans une même réaction. Elle comprend plusieurs étapes chronologiques.

II.1.2.1.1. Extraction de l'ADN bactérien

L'extraction d'ADN à été faite en utilisant la méthode de chauffage. A l'aide d'une pipette Pasteur (Berner, Helsinki), une masse de culture bactérienne provenant des colonies de 24 h a été prélevée et ajoutée dans un tube d'eppendorf (Hamburg, Allemangne) contenant 300 µl d'eau distillée stérile. Le tout a été placé dans un bain marie bouillant pendant 10 min. Puis le tube a été centrifugé à 11337 g pendant 10 min. Le surnageant a été utilisé pour la PCR.

II.1.2.1.2. Amplifiaction de l'ADN bactérien

- Amorces utilisées

Les amorces utilisées et leur concentration sont consignées dans le Tableau IV (annexe).

- Préparation du mix et programmes PCR

Les réactions PCR ont été conduites selon le protocole développé par Antikainen *et al.* (2009). Les mix et les programmes PCR ont été conformes aux indications de cet auteur. Le volume final de la réaction a été de 10 µl pour chaque échantillon et constitué de: 4,28 µl d'eau distillée stérile, 2 µl de tampon 5xGC (Finnzymes, F-519L, Espoo, Finlande), 0,4 µl de dNTPs (5 mM), 2 µl du mix des 16 paires d'amorces, 1 µl du surnageant contenant l'ADN bactérien, 0,24 µl de $MgCl_2$ (25 mM) 0,08 µl de la *Taq* polymérase (2U/ µl) (Finnzymes, F-540L, Espoo, Finlande). Le programme d'amplification dans le termocycleur (Bio-Rad Laboratories, Hercules, CA, USA) a été comme suit: 30 sec à 98°C, suivi de 35 cycles de 30 sec à 98°C, 60 sec à 63°C, 90 sec à 72°C, et une étape finale de 10 min à 72°C.

II.1.2.1.3. Electrophorèse sur gel d'agarose

Du gel d'agarose (Seakem ME Agarose for gel electrophoresis) à 1,5% (p/v) a été préparé dans du tampon 0,5×TBE (100 mM Trizma base, 100 mM acide borique, 2 mM EDTA) et coulé sur le support de gel pour l'électrophorèse. La cuve d'électrophorèse a été remplie avec le tampon 0,5×TBE jusqu'à immersion totale du gel. Les ADNs amplifiés par PCR ont été mélangés avec du tampon de charge (0,040% bromophénol, 7% glycérol, 6 mM EDTA), et 10 µl de ce mélange ont été prélevés à l'aide d'une micropipette et déposés dans des puits du gel d'agarose. Un marqueur de poids moléculaire de 100pb (GeneRuler, Fermentas) a été utilisé. La migration électrophorétique a été faite à 120 volts pendant 120 min. A la fin de la migration, le gel a été coloré dans une solution de bromure d'éthidium (BET, 10%) pendant 20 min. La visualisation a été faite sous une lampe à U.V. (Ultra-Violet) avec un imageur, intégrant le logiciel informatique Alpha Imager (Alpha Innotech Corporation, CA, USA) et connecté à une imprimante numérique monochrome de modèle P93DW (MITSUBISHI Electric, Malaysie) à l'aide de laquelle la photo du gel à été imprimée.

Les souches suivantes ont été utilisées comme références: RH 4283 (E 2348/69, [Baldini *et al.*, 1983]) pour EPEC, RH 3533 (ATCC 35401) pour ETEC, RH 4270 (ATCC 43895) pour STEC, RH 6647 (145-46-215, Statens Serum Institute [SSI], Copenhagen, Denmark) pour EIEC et IH 56822 (patient isolate, [Keskimäki *et al.*, 2000]) pour EAEC. Le contrôle négatif était *E. coli* RHE6715 (ATCC25922).

II.1.2.1.4. PCR simple pour la confirmation

Après les résultats de l'électrophorèse sur gel d'agarose, une PCR spécifique des différents gènes positifs a été effectuée pour confirmer la présence de chaque gène. Dans cette PCR, le mix a été constitué d'amorce spécifique pour chaque gène. Cette PCR a été nommée single-plex PCR.

II. 2. Sérotypage et antibiogramme des souches de *Salmonella*

- Sérotypage

Les souches pures de *Salmonella* conservées à 4°C dans des tubes contenant la gélose trypticase soja ont été ensemencées sur la gélose Drygalski (Oxoid, Basingstoke, Angleterre) et un dépôt centrale à été effectué sur la gélose Swarm à l'aide d'une pipette Pasteur stérile (Berner, Helsinki). Après 20 h d'incubation, la pureté des souches a été vérifiée sur la gélose Drygalski (Oxoid, Basingstoke, Angleterre) et leur mobilité sur la gélose Swarm (Oxoid, Basingstoke, Angleterre).

Pour la réalisation du sérotypage, une goutte d'antisérum a été déposée sur une plaque de verre stérile. Puis, un peu de culture bactérienne a été émulsionnée à l'anse de façon à obtenir un trouble homogène dans la goutte. Enfin, la plaque a été agitée par des mouvements circulaires jusqu'à l'apparition d'agglutinats.

Les colonies sur Drygalski (Oxoid, Basingstoke, Angleterre) ont servi à la détermination de l'antigène O et les colonies sur Swarm (Oxoid, Basingstoke, Angleterre) à la détermination de l'antigène H par agglutination des antisérums spécifiques selon le schéma de Kauffmann-White (Kauffmann, 1971).

L'identification sérologique des souches isolées de salmonelles a été faite d'abord par la determination du groupe en utlisant les antisérums O mélanges « OM » OMA, OMB, OMC, OMD, OME, OMF, OMG (Statens Serum Institute [SSI], Copenhagen, Denmark), suivie de l'identification du sérogroupe en utilisant les antisérums flagellaires HA, HB, HC, HD, HE et HF (Statens Serum Institute [SSI], Copenhagen, Denmark) selon le schéma de Kauffmann-White. Quelques structures antigéniques avec les différents antisérums sont présentées dans les tableaux III et IV de l'annexe 4.

- Antibiogramme

L'antibiogramme permet de mesurer la capacité d'un antibiotique à inhiber la croissance bactérienne *in vivo*. La sensibilité aux antibiotiques des souches de *Salmonella* a été réalisée en utilisant la méthode de diffusion à partir de disques imprégnés d'antibiotiques et la mesure des diamètres d'inhibition (CLSI, 2009). Les souches de référence suivantes, *E. coli* ATCC 25922, *E. coli* ATCC 35218, *Pseudomonas aeruginosa* ATCC 27853, *Enterococcus faecalis* ATCC 29212 et

Staphylococcus aureus ATCC 29213 ont été utilisées pour un contrôle de qualité et la validation des résultats.

Les antibiotiques utilisés dans cette étude sont cités ci-après selon leur ordre de disposition et de lecture sur la boite: Ampicilline (10 µg), Chloramphénicol (30 µg), Streptomycine (10 µg), Sulphonamides (300 µg), Trimethoprimes (5 µg), Ciprofloxacine (5 µg), Tétracycline (30 µg), Gentamicine (10 µg), Acide Nalidixique (30 µg), Cefotaxime (5 µg), Mécilliname (10 µg), Imipenème (10 µg).

III. Electrophorèse en champ pulsé des souches de *Salmonella* spp
III.1. Principe

Le principe de l'électrophorèse en champ pulsé consiste à alterner l'orientation du champ électrique au cours du temps. Chaque changement de champ électrique réoriente la molécule d'ADN dans le gel augmentant ainsi la probabilité que la molécule d'ADN soit orientée de façon à passer à travers les mailles du gel. Cette probabilité dépend de la taille de la molécule et la vitesse de migration d'un fragment d'ADN dans le gel varie dans le sens inverse de sa taille. L'électrophorèse en champ pulsé permet ainsi de séparer des fragments d'ADN d'une taille allant de moins de 1 kb à une dizaine de mégabases. Pour ce type d'électrophorèse, il n'est pas possible d'utiliser des ADN purifiés par les techniques classiques car ces techniques les cassent en fragments d'une taille inférieure à 100 kb. Pour éviter la cassure mécanique des molécules d'ADN les cellules sont incluses dans des blocs d'agarose.

III.2. Réalisation

Le protocole utilisé pour cette étude est celui décrit par PulseNet (2002). Il comporte trois grandes étapes subdivisées en plusieurs parties:

III.2.1. Culture et extraction de l'ADN bactérien
- Culture des bactéries et préparation de la suspension bactérienne

Avant l'extraction de l'ADN, les bactéries ont été mises en culture. Un premier ensemencement a été réalisé sur la gélose Drigalski (Bio Merieux, Marcy l'Etoile, France) afin de vérifier la pureté des souches. Après 24 h d'incubation à 37°C, une colonie de cette gélose a été ensuite repiquée sur la gélose Trypticase Soja (TSA) (Oxoid, Basingstoke, Angleterre) et incubée à 37°C pendant 18-20 h. A partir de ces boites, une suspension bactérienne de concentration 0,5- 0,55 MacFarland a été réalisée dans un tampon CSB (Cells Suspension Buffer) (EDTA, 100 mM + Tris-HCl, 100 mM, PH 8,0).

- **Préparation des blocs d'agarose**

Un volume de 400 µl de la suspension bactérienne de chaque échantillon a été prélevé et placé dans des tubes d'eppendorf. Ces tubes ont été placés dans un bloc à 50°C, puis 20 µl de la protéinase K (20 mg/ml) a été ajouté dans chaque tube d'eppendorf sans mélanger. Ensuite, 400 µl de gel d'agarose (1%) (Seakem Gold Agarose) préalablement préparé avec le tampon 1 × TE (10 mM Tris +1 mM EDTA, pH 8.0) a été ajouté dans chaque tube. Puis, 400 µl du mélange de chaque tube ont été prélevés et distribués dans des puits d'une moule de bloc d'agarose (plugs mold) dont le support est une plaque en verre au dessus de la glace. Après solidification, les blocs ont été placés à l'aide d'un scalpel dans des flacons de 50 ml contenant 5 ml de solution de lyse (EDTA, 50 mM + N-Lauryl-Sacrosine, 1% + Tris, 50 mM, PH, 8.0) et 25 ml de protéinase K (20 mg/ml). Les flacons ont été incubés dans un bain marie de 54°C pendant 1- 2 h sous agitation. Après 2 h d'incubation, les blocs d'agarose ont été rincés 2 fois à l'eau puis 4 fois au tampon 1 × TE, le temps entre chaque lavage a nécessité une incubation à 50°C pendant 15 min sous agitation. Enfin, les blocs d'agarose ont été stockés à 4°C pour la digestion enzymatique.

- **Digestion enzymatique**

Tout d'abord, des tubes d'eppendorf stériles ont été numérotés selon le nombre d'échantillons. Puis 200 µl de la solution de H-buffer (Tampon H) 1:10 ont été ajoutés dans chaque tube. Ensuite, sur une plaque en verre, 2-2,5 mm de blocs d'agarose ont été découpés à l'aide de scalpel et ajoutés dans les tubes. Après 15 min à température ambiante, les 200 µl de H-buffer ont été retirés à l'aide d'une micropipette et remplacés avec 200 autres µl de H-buffer. Enfin, 5 µl d'enzyme de restriction *Xba*I (10U/ µl) ou *Bln*I (10U/ µl) ont été ajoutés dans chaque tube. Les tubes ont été incubés à 37°C pendant 2 h.

III.2.2. Séparation électrophorétique et révélation des profils de macrorestrition

- Séparation électrophoretique

Après 2h d'incubation avec *Xba*I ou *Bln*I, la solution de H-buffer/*Xba*I ou H-buffer/*BlnI* a été retirée des tubes, et 200 µl de 0,5 × TBE (100 mM Trizma base, 100 mM acide borique, 2 mM EDTA) ont été ajoutés dans chaque tube et laissés à température ambiante pendant 5 min. Ensuite, les morceaux de bloc d'agarose (2-2,5 mm) ont été soigneusement retirés et déposés sur un peigne, puis laissés pour séchage pendant 5 min. Ensuite, le tout a été placé dans le bac avant de couler le gel (1%) préalablement préparé dans 0,5 × TBE. Après solidification, le peigne a été retiré et les puits ont été fermés avec le reste du gel. L'ensemble gel+blocs d'agarose solidifié a été placé dans la cuve d'électrophorèse qui contient préalablement 2 litres de 0,5 × TBE refroidis à 14°C à l'aide du refroidisseur de la machine d'électrophorèse. Les paramètres d'électrophorèse sont les suivant : Temps initial = 2 secondes; Temps final = 64 secondes; Durée de l'électrophorèse = 20 heures; Voltage = 6 V; Angle = 120°.

- Révélation des profils de macrorestrition

A la fin de la migration, le gel a été coloré dans une solution de bromure d'éthidium (BET, 10%) pendant 30-40 min. La visualisation a été faite sous une lampe à UV (Ultra- Violet) avec un imageur, intégrant le logiciel informatique Alpha Imager (Alpha Innotech Corporation, CA, USA) et connecté à une imprimante numérique monochrome de modèle P93DW (MITSUBISHI Electric, Malaysie) à l'aide de laquelle la photo du gel à été imprimée pour disposer d'un document archivable.

L'ADN de *Salmonella* Braenderup H9812 a été utilisé comme marqueur de poids moléculaire pour la comparaison des souches analysées dans différents gels.

L'ADN de certains sérotypes, de certains clones de salmonelles se dégrade en présence du Tris. Ces souches ont été ré-analysée avec un autre protocole d'extraction et de migration où le Tris présent dans différents tampons a été remplacé par de l'hepes et du thiourée à la concentration finale de 100 µM (Liesegang et Tschäpe, 2002).

III.3. Analyse des résultats

L'analyse de l'ensemble des résultats a été effectuée à l'aide du logiciel BioNumerics (Biorad, Applied-Maths, Sint-Martens-Latem, Belgique). Ce logiciel permet de créer et gérer des bases de données ainsi que d'analyser de nombreuses expériences. Chaque souche correspond à une entrée dans la base de données et peut être associée à une ou plusieurs expériences. Les résultats obtenus lors des différentes expériences ou des analyses effectuées permettent d'estimer un lien génétique entre les différents isolats. Un numéro de profil est attribué suite à l'analyse des résultats de PFGE, il est propre à la base de données de l'unité Caractérisation et Epidémiologie des Bactéries (CEB) où ont été réalisés les typages mais a été défini selon une nomenclature européenne par un code à 6 lettres associé à un numéro de 4 chiffres (Peters *et al.*, 2003). La première lettre correspond au genre (S pour *Salmonella*), les 3 lettres suivantes correspondent à la codification du sérovar (ENT pour Enteritidis). Les 2 dernières lettres correspondent au code de l'enzyme (XB pour *Xba*I). Par exemple le numéro SENTXB0001 correspond au premier profil observé de *Salmonella* Enteritidis avec l'enzyme *Xba*I. Ce logiciel peut effectuer diverses analyses telles que le calcul de dendrogramme pour visualiser les groupes d'individus semblables. Il se calcule en multipliant par deux le nombre de fragments communs, divisé par le nombre total de fragments dans les deux profils. Différents paramètres doivent être pris en compte pour l'identification des fragments (l'intensité minimale lors de la recherche automatique, l'élimination des artefacts, l'addition de doubles bandes) et pour la reconnaissance d'homologie de position (degré de tolérance des erreurs résiduelles d'alignement et de position de fragments). Ainsi, certains profils qui présentent des différences minimes peuvent avoir le même numéro. Après la comparaison deux à deux de tous les profils, une matrice de similarité est obtenue. Le logiciel est capable de construire une représentation graphique de ces matrices sous forme de dendrogramme. Nous avons utilisé la méthode UPGMA (Unweighted Pair

Group Method Using Arithmetic Averages) basée sur le regroupement suivant la distance moyenne. Cette méthode utilise un algorithme de regroupement séquentiel dans lequel les relations sont identifiées dans l'ordre de leur similarité et la reconstruction de l'arbre suite à cet ordre. Il y a d'abord l'identification des deux séquences les plus proches et ce groupe est ensuite traité comme un tout, puis la séquence la plus proche est recherchée, ainsi de suite jusqu'à ce qu'il n'y ait plus que deux groupes.

TROISIEME PARTIE

RESULTATS ET DISCUSSION

Chapitre 1: Etat sanitaire des viandes vendues sur les marchés publiques de Ouagadougou

I. Prévalence de *Salmonella* et *E. coli* sur les viandes de bœuf, mouton et poulet

Les infections à *Salmonella* et *Escherichia coli* posent un problème de santé publique à travers le monde. La plupart de ces infections sont d'origine alimentaire bien que certaines peuvent être acquises par contact direct avec les animaux. Ces derniers sont les principaux réservoirs de *Salmonella* et *E. coli* qui peuvent contaminer les viandes au cours de l'abattage si les conditions d'hygiènes ne sont pas respectées. Au Burkina Faso, il y a très peu de données publiées sur la prévalence dans les aliments de ces bactéries zoonotiques qui constituent une menace pour la santé des consommateurs. Ainsi, nous avons mené une étude sur les viandes de grande consommation dans la ville de Ouagadougou pour déterminer la prévalence de *Salmonella* et *Escherichia coli*.

Des échantillons de viande de bœuf, de mouton et de boyaux de bœuf destinés à la consommation, ainsi que des carcasses entières de poulets ont été prélevés dans des conditions naturelles de vente pour des analyses microbiologiques. Les souches de *Salmonella* isolées ont été caractérisées sur le plan biochimique et sérotypées après confirmation par la galerie API 20 E. La sensibilité à 12 antibiotiques de ces bactéries a été également testée. La présence des *E. coli* indicateurs de contamination fécale, a été détectée par les méthodes standards.

La prévalence globale de *Salmonella* a été de 9,3% sur les viandes avec six sérotypes différents (Agona, Bredeney, Derby, Hato, Tilene et Senftenberg) rapportés pour la première fois au Burkina Faso. La résistance aux antibiotiques a été constatée seulement chez les sérotypes Derby. La prévalence totale de *E. coli* indicateurs de contamination fécale a été de 100%. L'enquête menée sur les bouchers nous a informés sur les mauvaises conditions d'hygiène lors de la vente et du transport des viandes, le manque d'installation sanitaire dans les boucheries, la vente à température ambiante sans aucune protection, le transport par différents moyens allant de la bicyclette aux véhicules non climatisés vers les marchés. Ces résultats ont fait l'objet d'un article publié (**Article 1**).

II. Article 1: Prevalence of *Salmonella enterica* **and the Hygienic Indicator** *Escherichia coli* **in Raw Meat at Markets in Ouagadougou, Burkina Faso. [***Journal of Food Protection***, 74: 1547-1551 (2011)].**

1547

Journal of Food Protection, Vol. 74, No. 9, 2011, Pages 1547–1551
doi:10.4315/0362-028X.JFP-11-124
Copyright ©, International Association for Food Protection

Research Note

Prevalence of *Salmonella enterica* and the Hygienic Indicator *Escherichia coli* in Raw Meat at Markets in Ouagadougou, Burkina Faso

ASSÈTA KAGAMBÈGA,[1,2] KAISA HAUKKA,[2*] ANJA SIITONEN,[2] ALFRED S. TRAORÉ,[1] AND NICOLAS BARRO[1]

[1]Centre de Recherches en Sciences Biologiques, Alimentaires et Nutritionnelles, Département de Biochimie-Microbiologie, UFR des Sciences de la Vie et de la Terre, Université de Ouagadougou, 03 B.P. 7021 Ouagadougou 03, Burkina Faso; and [2]Bacteriology Unit, National Institute for Health and Welfare (THL), P.O. Box 30, 00271 Helsinki, Finland

MS 11-124: Received 15 March 2011/Accepted 6 May 2011

ABSTRACT

This study investigated the hygienic status and prevalence of *Salmonella* and *Escherichia coli* in retail meat sold at open markets in Ouagadougou, Burkina Faso. A total of 150 samples of beef meat ($n = 45$), beef intestine ($n = 45$), mutton ($n = 30$), and chicken ($n = 30$) were collected from four local markets for investigation. The prevalence of *Salmonella enterica* subsp. *enterica* was 9.3%, and six serotypes, all previously unreported in Burkina Faso, were identified: Derby, Tilene, Hato, Bredeney, Agona, and Senftenberg. Most of the *Salmonella* isolates were sensitive to the 12 antimicrobial drugs tested. The prevalence of *E. coli* was 100% in all the meat types. An assessment of hygiene practices for the production, transportation, display, and vending of the meat revealed unhygienic conditions. Meat sellers had a low education level and poor knowledge of foodborne pathogens and their transmission routes. The findings showed that foodstuff handlers were in dire need of education about safe food handling practices.

Foodborne illnesses are a worldwide public health concern and a significant cause of reduced economic growth *(40)*. These illnesses can result from consumption of food contaminated with pathogenic organisms or toxic chemicals. The main symptoms of foodborne infection and/or intoxication include nausea, vomiting, colic, and diarrhea. Different organisms can cause foodborne infections, the most common bacterial agents being *Salmonella*, *Campylobacter*, and diarrheagenic *Escherichia coli* *(30, 32, 35)*. Each year, *Salmonella* and *E. coli* O157:H7 are estimated to each cause, respectively, 1.3 million and 62,000 cases of gastroenteritis in the United States *(21)*. Also in developing countries, these bacteria are a public health concern and are a common source of medical complications, especially in human immunodeficiency virus–infected patients *(22, 26)*.

Contaminated raw meat and meat products are considered major vehicles for transmitting foodborne pathogens to humans *(20, 23, 30, 33)*. Freshly slaughtered animals may harbor relatively few bacteria, but the meat surface is exposed to contamination during slaughter, evisceration, and other operations after slaughter *(28)*. Transportation conditions and exposure during vending operations can lead to contamination *(3, 4)*. Reducing meat contamination will reduce the risk of transmission of pathogenic bacteria and foodborne diseases to consumers.

* Author for correspondence. Tel: +358-206108172; Fax: +358-206108238; E-mail: kaisa.haukka@thl.fi.

In some developing countries, low socioeconomic level, poverty, and low educational levels lead to deficiencies in awareness, hygienic and sanitation facilities, rules, quality control, and alert systems, meaning foodborne infections are a persistent burden *(5, 24)*.

This study was initiated to determine the prevalence of *Salmonella enterica* subsp. *enterica* and *E. coli* in raw meat bought from the local markets in Ouagadougou, Burkina Faso. Also *Salmonella* serotypes and their antimicrobial sensitivity were determined. In addition, the hygienic practices of the meat sellers were investigated.

MATERIALS AND METHODS

Study design. The study was carried out between October 2008 and February 2009, that is, during the dry season with average temperature between 30 and 38°C. Ten meat-selling stalls at four large open markets located in low socioeconomic areas of Ouagadougou were enrolled into the study. They were each visited 15 times during the study. Prior to any other investigations, the aims of the study were explained to the meat sellers. They were recruited into the study after the owners of the vending places had given their consent and the sellers were assured of confidentiality in accordance with the research protocol approved by the ethical committee. A questionnaire used by Angelillo et al. *(2)* was modified and adapted to this study. Completion of the questionnaire was followed by a face-to-face interview in the local language. Observations of the working methods of the meat sellers and their stall surroundings were made, as well as of their hygienic practices, with instances of unhygienic behavior recorded.

J. Food Prot., Vol. 74, No. 9

Sampling. A total of 45 beef meat, 45 beef intestine, 30 mutton, and 30 chicken samples were collected. All the meat samples, except the chicken, came from animals freshly slaughtered in a national slaughterhouse. Beef intestines were cleaned with water in a slaughterhouse and sold by the meat sellers for human consumption. Chickens were killed at the market sites. About 200 g of beef, beef intestine, mutton, or whole chickens were purchased from the four local markets and placed in sterile plastic bags. The samples were transported to the laboratory on ice and processed within 2 h of sampling.

Processing of samples and microbial identification. For *Salmonella* isolation, 25 g of a meat sample was added to 225 ml of buffered peptone water (BPW; Liofilchem, Teramo, Italy) and homogenized in a stomacher lab-blender 400 (Seward, London, England). After a 24-h incubation at 37°C, 2 ml of the preenrichment broth was inoculated to 10 ml of selenite-cystine broth (Oxoid, Basingstoke, England), which was incubated at 37°C for 24 h. Two loopfuls were streaked onto Hektoen agar (Oxoid) plates and incubated at 37°C for 24 h.

Fifty homogenized samples in BPW were also streaked onto Mueller-Hinton agar and incubated at 37°C for 24 h. The bacterial mass was collected and streaked onto xylose lysine deoxycholate agar (XLD; Oxoid) and incubated at 37°C for 24 h.

The identity of typical-looking *Salmonella* colonies on Hektoen and XLD was examined by using orthonitrophenyl-β-D-galactopyranoside (ONPG), citrate, mannitol, lysine decarboxylase tests, and the Kliger Hajna medium (Liofilchem). Finally the isolates were confirmed by API 20E (BioMérieux, Marcy l'Etoile, France). All *Salmonella* isolates were verified and typed further by the *Salmonella* Reference Laboratory at the National Institute for Health and Welfare, Helsinki, Finland. The susceptibility of strains to 12 antimicrobials was evaluated by the disk diffusion method on Mueller-Hinton agar (Oxoid) at 37°C for 24 h. The antimicrobials evaluated were ampicillin, chloramphenicol, streptomycin, sulfonamide, tetracycline, trimethoprim, ciprofloxacin, nalidixic acid, mecillinam, imipenem, neomycin, and ceftriaxone.

For *E. coli* detection, a loopful of the homogenized sample in BPW was streaked onto eosin methylene blue agar (Difco, BD, Sparks, MD) plates and incubated at 37°C for 18 h. The plates were inspected for presumptive *E. coli* colonies with a metallic-green color and a dark or purple center. The suspected *E. coli* colonies were streaked onto Mueller-Hinton agar. Typical *E. coli* colonies were confirmed by negative urease and positive indole tests.

RESULTS

Description of carcass transportation and vending conditions. Transportation systems such as a bicycle, a motorcycle, or a car without refrigeration were used for distributing beef and mutton carcasses from a slaughterhouse to the local markets. Carcasses were only partially wrapped or not wrapped at all during transport. At the market, the meat sellers carried the carcasses on their shoulders. Carcasses were stored by hanging, and meat was sold off a table at ambient temperature without protection from dust and flies at any point during the day.

The slaughter of chickens was done by the traditional slaughtering method at the market sites. Killing was done by the butchers by hand. All the subsequent operations such as bleeding, plucking, evisceration, and cutting were executed on the same table. After that, the chickens were rinsed in the

TABLE 1. *Characteristics of meat sellers and the vending environment*

Characteristics	No. (%)
Meat sellers ($n = 31$)	
Male	31 (100)
No. of persons per stall	2–4
Answer declined[a]	9 (29)
Wearing overall	7 (23)
Age range (yr)	12–50
Educational level ($n = 22$)	
Primary or less	17 (77)
Secondary	5 (23)
Knowledge of foodborne pathogens ($n = 22$)	
Yes	6 (27)
No	16 (73)
Stall characteristics ($n = 10$)	
Modern building	2 (20)
Traditional stall	8 (80)
Tap water facilities	2 (20)
Presence of a garbage bin	0
Presence of liquid waste	10 (100)
Presence of solid waste	10 (100)
Presence of animals (vultures, dogs)	10 (100)
Protection of meat during vending	0
Presence of meat and giblets on the same table	10 (100)
Use of the same knife for meat and giblets	10 (100)

[a] Declined to answer questions regarding their age, educational level, and knowledge of foodborne pathogens.

same water in a large bowl and displayed at ambient temperature on a large aluminum plate without protection.

All meat sellers were males and generally persons without any knowledge of modern meat processing (Table 1). Seventy-three percent of the meat sellers were unaware that foodborne pathogens can be transmitted by meat. Nine (29%) of the 31 meat sellers declined the interview, although a local language was used. Most of the meat stalls were traditional stalls covered with steel sheet or by straw. Their selling environment was infested by pets, vultures, flies, and lizards. After the market was closed, these animals could be seen on or roaming around the vending tables, licking or sucking traces of blood or meat residues. Potable water facilities were absent from several stalls. The next day, the selling tables were not cleaned adequately before reception of new carcasses for sale.

***Salmonella* and *E. coli* prevalence on meat and intestine samples.** Raw meat samples were frequently contaminated by enteropathogenic bacteria (Table 2). A total of 14 *Salmonella* isolates were obtained from Hektoen plates; thus, the general prevalence of *Salmonella* was 9.3% in all samples (14 of 150). *Salmonella* was not found in beef but in 13% (6 of 45) of beef intestine, in 7% (2 of 30) of mutton, and in 37% (11 of 30) of chicken samples. In addition, the XLD plates were tested for *Salmonella* isolation, and they yielded five further isolates. *E. coli* was found in all of the 150 meat samples analyzed.

TABLE 2. *Prevalence of* Salmonella *and* E. coli *on raw meats and intestines sold at open markets in Ouagadougou*

Meat type (n = 150)	No. (%) of samples from which bacteria were isolated	
	Salmonella[a]	E. coli
Beef (n = 45)	0	45 (100)
Beef intestine (n = 45)	6 (13)	45 (100)
Mutton (n = 30)	2 (7)	30 (100)
Chicken (n = 30)	11 (37)	30 (100)
Total prevalence	14 (9.3)	150 (100)

[a] Isolates from Hektoen agar. In addition, three isolates from beef intestine, one from mutton, and one from chicken samples were obtained with XLD agar.

Salmonella serotypes and their antimicrobial sensitivity. Among the 19 *Salmonella* isolates, Derby (9 strains, 42%) and Tilene (5 strains, 26%) were the most common serotypes. The other serotypes detected were Senftenberg, Bredeney, Agona, and Hato. Five serotypes (Senftenberg, Bredeney, Agona, Tilene, and Hato) were found in the beef intestine samples, three serotypes (Derby, Agona, Tilene) in the chicken samples, and one serotype (Derby) in the mutton samples.

Three *Salmonella* Derby isolates were resistant to tetracycline, streptomycin, and sulfonamide, and nine other *Salmonella* isolates had intermediate resistance to sulfonamide.

DISCUSSION

Our study showed that the hygienic practices of the meat sellers at the four studied markets of Ouagadougou did not meet the hygiene levels for the handling of meat products as recommended by World Health Organization and the Food and Agriculture Organization joint committee *(12)*. Indeed, the sellers never stored meat at an appropriate low temperature or protected it against flies; they never washed their hands, wore protective clothing, or used adequate amounts of water to clean their tables and cutting tools. It is evident that these conditions can lead to contamination of meat and to various cross-contamination situations, as has been previously reported *(1, 4, 38)*. In our study, only one-quarter of meat sellers had any knowledge of the transmission routes of enteropathogenic bacteria to humans and only one-fifth of the stalls had water facilities available.

Microbiological investigation of the meat samples revealed that they were contaminated with enteropathogenic bacteria. This finding raised three main questions: (i) what is the rate of contamination for enteric bacteria, (ii) when does the contamination happen, and (iii) what is the risk to consumer health? *E. coli* is an indicator of fecal contamination, while certain particular strains of these bacteria are pathogenic, for example, *E. coli* O157. According to our study, all the meat samples from the markets of Ouagadougou were contaminated by *E. coli*. For comparison, a study carried out in Pretoria, South Africa, revealed a lower contamination rate with *E. coli*, of 74.5%, in meat and meat

products *(39)*. In Mexico, the *E. coli* contamination rate was 43% *(13)*. A study in the United States showed a prevalence of *E. coli* of 38.7% in chicken meat, 19.0% in beef meat, and 16.3% in pork meat *(43)*. In Australia, *E. coli* was isolated from 17.8% of ground beef and 16.7% of diced lamb samples *(27)*. The differences in contamination levels are probably due to the national or geographic characteristics of animal feeding systems, differences in meat processing environments, different testing methodologies, and different frequency and quantity of sample testing *(9, 14, 16, 17, 29)*.

In the present study, the prevalence of *Salmonella* contamination varied from 0% in beef meat to 37% in chicken carcasses purchased from the markets. Several studies have shown that chicken and cattle are reservoirs for *Salmonella (8, 10, 15, 18, 25, 31)*. *Salmonella* infection in these animals can be transmitted to humans by several routes such as the environment, many of which are difficult to control *(18)*. Highly varying levels of *Salmonella* contamination in meat have been reported in the literature. For example, 8% prevalence of *Salmonella* in beef was found in Ireland *(19)* and as high as 90% prevalence in beef in Senegal *(34)*. Four percent of the mutton samples were found to contain *Salmonella* in New Zealand *(41)*, whereas the occurrence of *Salmonella* in chicken carcasses was reported to be 19% in a study conducted in South Africa *(37)*. As with *E. coli*, the differences in sample collection and analyses can explain some of the observed differences in prevalence of *Salmonella*. In our investigation Hektoen agar was used throughout the study, though XLD agar was also utilized. Our experience was that *Salmonella* grew better on the XLD agar, and so we recommend it for future studies.

Curiously, the most common *Salmonella* serotypes worldwide, Enteritidis and Typhimurium, were not found in this study. Instead, six less common serotypes such as Derby, Tilene, Agona, Senftenberg, Bredeney, and Hato were identified among the 19 *Salmonella* isolates. *Salmonella* Derby was found in mutton and chicken samples. This serotype has earlier been reported to be one of the most common serotypes isolated from swine *(36)*. The second most common serotype in our study was *Salmonella* Tilene. Several studies have indicated its origin to be pygmy hedgehogs *(36, 42)*. The repeated finding of Tilene in this study might thus be explained by the nature of animal husbandry in Burkina Faso. Traditionally, cows and sheep roam freely at pasture in the bush. The wild animals living in such places, such as hedgehogs, can contaminate grasses with their excreta. However, thus far we have not detected this serotype in our human feces samples (unpublished results). *Salmonella* Agona was found in three meat samples. This serotype is commonly recovered from domestic animals and also from human feces *(11)*.

Public health concerns about foodborne pathogens extend to antimicrobial resistance. Most of our *Salmonella* isolates were sensitive to the antimicrobial drugs commonly used in Burkina Faso. Only three of the *Salmonella* Derby strains showed resistance to commonly used tetracycline, streptomycin, and sulfonamide.

In many Western countries, it is required that state authorities be notified of infections caused by *Salmonella*. The surveillance and typing (serotyping, phage typing, and susceptibility patterns) of *Salmonella* strains in developing countries would allow the detection of emerging new strains as well as antimicrobial resistance. This information would help in formulating prevention strategies.

In conclusion, our study showed that *Salmonella* and *E. coli* occur commonly in beef, intestines, mutton, and chicken sold to consumers in Burkina Faso. The presence of *Salmonella* and the high prevalence of *E. coli* in meat are a threat to consumers' health and warrant serious attention. The meat sellers included in this study were limited to 10 stalls at four open markets in Ouagadougou, but they represented very typical meat selling sites in Burkina Faso and other developing countries. Although cooking times and temperatures for meat prepared for human consumption may be adequate to eliminate bacteria on the surface of meat, they are not necessarily sufficient to kill bacteria that may have invaded inside the meat (7). Also, cross-contamination by contact with surfaces previously contaminated with raw meat and contact with the raw meat itself, with utensils or hands, can also lead to a health risk for consumers, even if raw foodstuff is always cooked before consumption (6, 13). Our data highlighted the transmission route of pathogenic bacteria from farms, via animals, meat sellers, and the vending environment to raw meat sold at the markets in Ouagadougou, Burkina Faso. The transmission route may also lead to an infection risk for consumers. This study demonstrated that there is a need to increase awareness of environmental and personal hygiene issues among the Burkinabe population, especially among food vendors.

ACKNOWLEDGMENTS

This was a collaborative study between the Finnish National Institute for Health and Welfare (THL) and CRSBAN and the University of Ouagadougou that was funded by the Academy of Finland grant 122600. We thank the personnel of the Enteric Bacteria Laboratory at THL for help in verifying and serotyping the isolates. We also thank all of the meat sellers involved in the study.

REFERENCES

1. Altekruse, S. F., D. A. Street, S. B. Fein, and A. S. Levy. 1996. Consumer knowledge of foodborne microbial hazards and food-handling practices. *J. Food Prot.* 59:287–294.

2. Angelillo, I. F., M. R. Foresta, C. Scozzafava, and M. Pavia. 2001. Consumers and foodborne diseases: knowledge, attitudes and reported behavior in one region of Italy. *Int. J. Food Microbiol.* 64: 161–166.

3. Arsenault, J., A. Letellier, S. Quessy, and M. Boulianne. 2007. Prevalence and risk factors for *Salmonella* and *Campylobacter* spp. carcass contamination in broiler chickens slaughtered in Quebec, Canada. *J. Food Prot.* 70:1820–1828.

4. Barro, N., A. R. Bello, A. Savadogo, C. A. T. Ouattara, A. J. Ilboudo, and A. S. Traoré. 2006. Hygienic status assessment of dishwaters, utensils, hands and pieces of money in street foods vending sites in Ouagadougou, Burkina Faso. *Afr. J. Biotechnol.* 5:1107–1112.

5. Barro, N., A. A. Gamene, Y. Itsiembou, A. Savadogo, A. P. Nikiema, C. A. T. Ouattara, C. A. De Souza, and A. S. Traoré. 2007. Street-vended foods improvement: contamination mechanisms and application of food safety objective strategy: critical review. *Pak. J. Nutr.* 6: 1–10.

6. Bryan, F. L. 1988. Risk associated with practices, procedures, and processes that lead to outbreaks of foodborne diseases. *J. Food Prot.* 51:663–673.

7. Cardinale, E., J. D. Perrier Gros-Claude, F. Tall, E. F. Gueye, and G. Salvat. 2005. Risk factors for contamination of ready-to-eat street-vended poultry dishes in Dakar, Senegal. *Int. J. Food Microbiol.* 103: 157–165.

8. Cardinale, E., F. Tall, E. F. Guèye, M. Cissé, and G. Salvat. 2004. Risk factors for *Salmonella enterica* subsp. *enterica* infection in Senegalese broiler-chicken flocks. *Prev. Vet. Med.* 63:151–161.

9. Chahed, A. 2008. Prévalence et caractérisation de souches d'*Escherichia coli* O157 producteurs de shiga-toxines isolées de denrées alimentaires d'origine animale en Belgique et en Algérie. *Med. Vet.* 152:39–43.

10. Chambers, J. R., J. R. Bisaillon, Y. Labbe, C. Poppe, and C. F. Langford. 1998. *Salmonella* prevalence in crops of Ontario and Quebec broiler chickens at slaughter. *Poult. Sci.* 77:1497–1501.

11. Clark, G. M., A. F. Kaufmann, E. J. Gangarosa, and M. A. Thompson. 1973. Epidemiology of an international outbreak of *Salmonella* Agona. *Lancet* ii:490–493.

12. Codex Alimentarius Commission. 2005. Code of hygienic practice for meat. Adopted by the CAC, New Zealand, July 2005 (CAC/RCP 58–2005). Codex Alimentarius Commission, Geneva.

13. Estrada-Garcia, T., C. Lopez-Saucedo, B. Zamarripa-Ayala, M. R. Thompson, L. Gutierez-Cogo, I. Perez-Martinez, and A. Escobar-Gutierrez. 2004. Prevalence of *Escherichia coli* and *Salmonella* spp. in street-vended food of open markets (tianguis) and general hygienic and trading practices in Mexico City. *Epidemiol. Infect.* 132:1181–1184.

14. Fratamico, P. M., A. A. Bhagwat, L. Injaian, and P. J. Fedorka-Cray. 2008. Characterization of Shiga toxin-producing *Escherichia coli* strains isolated from swine feces. *Foodborne Pathog. Dis.* 5:827–838.

15. Hald, T., D. Vose, H. C. Wegener, and T. Koupeev. 2004. A Bayesian approach to quantify the contribution of animal-food sources to human salmonellosis. *Risk Anal.* 24:255–269.

16. Hussein, H. S., and T. Sakuma. 2005. Prevalence of Shiga toxin producing *Escherichia coli* in dairy cattle and their products. *J. Dairy Sci.* 88:450–465.

17. Kegode, R. B., D. K. Doetkott, M. L. Khaitsa, and I. V. Wesley. 2008. Occurrence of *Campylobacter* species, *Salmonella* species and generic *Escherichia coli* in meat products from retail outlets in the Fargo metropolitan area. *J. Food Saf.* 28:111–125.

18. Lindqvist, N., A. Siitonen, and S. Pelkonen. 2002. Molecular follow-up of *Salmonella enterica* subsp. *enterica* serovar Agona infection in cattle and humans. *J. Clin. Microbiol.* 40:3648–3653.

19. McEvoy, J. M., A. M. Doherty, J. J. Sheridan, I. S. Blair, and D. A. McDowell. 2003. The prevalence of *Salmonella* spp. in bovine faecal, rumen and carcass samples at a commercial abattoir. *J. Appl. Microbiol.* 94:693–700.

20. Mead, G. C. 1993. Problems of producing safe poultry: discussion paper. *Royal Soc. Med. J.* 85:39–42.

21. Mead, P. S., L. Slutsker, V. Dietz, L. F. McCaig, J. S. Bresee, C. Shapiro, P. M. Griffin, and R. V. Tauxe. 1999. Food-related illness and death in the United States. *Emerg. Infect. Dis.* 5:607–625.

22. Medeiros, M. I., S. N. Neme, P. da Silva, D. M. Capunno, M. C. Errera, S. A. Fernandes, G. R. do Valle, and F. A. de Avila. 2001. Etiology of acute diarrhea among children in Ribeiro Preto-SP, Brazil. *Rev. Inst. Med. Trop. Sao Paulo* 43:21–24.

23. Mensah, P., M. Armar-Klemesu, A. S. Hammond, A. Haruna, and R. Nyarko. 2001. Bacterial contaminants in lettuce, tomatoes, beef and goat meat from metropolitan Accra. *Ghana Med. J.* 35:1–6.

24. Mensah, P., D. Yeboah-Manu, K. Owusu-Darko, and A. Ablordey. 2002. Street foods in Accra, Ghana: how safe are they? *Bull. WHO* 80:546–554.

25. Mochizuki, Y., H. Masuda, S. Kanazashi, Y. Hosoki, K. Itoh, K. Ohishi, T. Nishina, Y. Handa, K. Shiozawa, and Y. Miwa. 1992. Clinical and epidemiological aspects of enteritis due to *Salmonella* Hadar. I. Isolation of *S. hadar* from sporadic diarrhea-clinical and bacteriological study. *J. Jpn. Assoc. Infect. Dis.* 66:22–29.

26. Obi, C. L., and P. O. Bessong. 2002. Diarrhoeagenic bacterial pathogens in HIV-positive patients with diarrhoea in rural communities of Limpopo Province, South Africa. *J. Health Popul. Nutr.* 20: 230–234.

27. Phillips, D., D. Jordan, S. Morris, I. Jenson, and J. Sumner. 2008. A national survey of the microbiological quality of retail raw meats in Australia. *J. Food Prot.* 71:1232–1236.

28. Price, J. F., and B. Schweigert. 1971. The service of meat and meat products, 2nd ed. W. H. Freeman and Company, San Francisco.

29. Rigobelo, E. C., E. Santo, and J. M. Marin. 2008. Beef carcass contamination by shiga toxin–producing *Escherichia coli* strains in an abattoir in Brazil: characterization and resistance to antimicrobial drugs. *Foodborne Pathog. Dis.* 5:811–817.

30. Roels, T. H., P. A. Frazak, J. J. Kazmierczak, W. R. Mackenzie, M. E. Proctor, T. A. Kurzynski, and J. P. Davis. 1997. Incomplete sanitation of meat grinder and ingestion of raw ground beef: contributing factors to a large outbreak of *Salmonella* Typhimurium infection. *Epidemiol. Infect.* 119:127–134.

31. Rose, N., F. Beaudeau, P. Drouin, J. Y. Toux, V. Rose, and P. Colin. 1999. Risk factors for *Salmonella enterica* subsp. *enterica* contamination in French broiler-chicken flocks at the end of the rearing period. *Prev. Vet. Med.* 39:265–277.

32. Sofos, J. N., S. L. Kochevar, G. R. Bellinger, D. R. Buege, D. D. Hancock, S. C. Ingham, J. B. Morgan, J. O. Reagan, and G. C. Smith. 1999. Source and extent of microbiological contamination of beef carcasses in seven United States slaughtering plants. *J. Food Prot.* 62:140–145.

33. Stern, N. J., P. Fedorka-Cray, J. S. Bailey, N. A. Cox, S. E. Craven, K. L. Hiett, M. T. Musgrove, S. Ladely, D. Cosby, and G. C. Mead. 2001. Distribution of *Campylobacter* spp. in selected U.S. poultry production and processing operations. *J. Food Prot.* 64:1705–1710.

34. Stevens, A., Y. Kaboré, J. D. Perrier-Gros-Claude, Y. Millemann, A. Brisabois, M. Catteau, J. F. Cavin, and B. Dufour. 2006. Prevalence and antibiotic-resistance of *Salmonella* isolated from beef sampled from the slaughterhouse and from retailers in Dakar (Senegal). *Int. J. Food Microbiol.* 110:178–186.

35. Tuttle, J., T. Gomez, M. P. Doyle, J. G. Wells, T. Zhao, R. V. Tauxe, and P. M. Griffin. 1999. Lessons from a large outbreak of *Escherichia coli* O157:H7 infections: insights into the infectious dose and method of widespread contamination of hamburger patties. *Epidemiol. Infect.* 122:185–197.

36. Valdezate, S., A. Vidal, S. Herrera-León, J. Pozo, P. Rubio, M. A. Usera, A. Carvajal, and M. A. Echeita. 2005. *Salmonella* Derby clonal spread from pork. *Emerg. Infect. Dis.*11:694–698.

37. van Nierop, W., A. G. Dusé, E. Marais, N. Aithma, N. Thothobolo, M. Kassel, R. Stewart, A. Potgieter, B. Fernandes, J. S. Galpin, and S. F. Bloomfield. 2005. Contamination of chicken carcasses in Gauteng, South Africa, by *Salmonella*, *Listeria monocytogenes* and *Campylobacter*. *Int. J. Food Microbiol.* 99:1–6.

38. Vollaard, A. M., S. Ali, H. A. van Asten, I. S. Ismid, S. Widjaja, L. G. Visser, C. Surjad, and J. T. van Dissel. 2004. Risk factors for transmission of foodborne illness in restaurants and street vendors in Jakarta, Indonesia. *Epidemiol. Infect.* 132:863–872.

39. Vorster, S. M., R. P. Greebe, and G. L. Nortje. 1994. Incidence of *Staphylococcus aureus* and *Escherichia coli* in ground beef, broilers and processed meats in Pretoria, South Africa. *J. Food Prot.* 57:305–310.

40. White, P. L., A. R. Baker, and W. O. James. 1997. Strategies to control *Salmonella* and *Campylobacter* in raw poultry products. *Rev. Sci. Tech.* 16:525–541.

41. Wong, T. L., C. Nicol, R. Cook, and S. MacDiarmid. 2007. *Salmonella* in uncooked retail meats in New Zealand. *J. Food Prot.* 70:1360–1365.

42. Woodward, D. L., R. Khakhria, and W. M. Johnson. 1997. Human salmonellosis associated with exotic pets. *J. Clin. Microbiol.* 35: 2786–2790.

43. Zhao, C., B. Ge, J. De Villena, R. Sudler, E. Yeh, S. Zhao, D. G. White, D. Wagner, and J. Meng. 2001. Prevalence of *Campylobacter* spp., *Escherichia coli*, and *Salmonella* serovars in retail chicken, turkey, pork, and beef from the Greater Washington, D.C., area. *Appl. Environ. Microbiol.* 67:5431–5436.

III. Caractérisation moléculaire des *E. coli* isolées sur la viande

Dans les pays en voie de développement, il est difficile de caractériser les pathovars de *E. coli*, étant donné la complexité et le coût des méthodes de caractérisation. Pourtant, la prévalence des *E. coli* indicateurs de contamination fécale, étant de 100% sur tous les types de viandes précedemment analysés, nous avons tenu à identifier les différents pathovars de *E. coli* qui pourraient être présents sur ces viandes. Pour cela, les *E. coli* indicateurs de contamination fécale isolés des viandes ont été soumis à une PCR multiplexe appelé la "16-plex PCR", développée par Antikainen *et al.* (2009) permettant la détection simultanée des 16 gènes de virulence des cinq principaux pathovars de *E. coli* responsables de différentes formes de diarrhées chez l'homme. Ces principaux pathovars sont: *E. coli* entéropathogènes (EPEC), *E. coli* entérotoxinogènes (ETEC), *E. coli* producteurs de shiga-toxines (STEC), *E. coli* entéroinvasifs (EIEC) et *E. coli* entéroaggrégatifs (EAEC). Cette étude nous a révélé une forte prévalence des pathovars de *E. coli* sur les différents échantillons de viande analysés, 44% sur les viandes de bœuf, 53% sur les boyaux de bœufs, 38% sur les viandes de mouton et 29% sur les carcasses de poulet. Parmi ces pathovars, STEC a été le plus prévalent avec une prédominance sur les boyaux. Le pathovar EIEC n'a été détecté sur aucune viande ou boyau. En ce qui concerne les différents gènes de virulence, la PCR a révélée une prévalence des gènes responsables de production de shiga-toxine (*stx*) 1 ou 2 de 28% dans les échantillons, qui sont des gènes de virulence de STEC. Les gènes de virulence de EPEC étaient *eae*, *esc*V et/ou *ent* et/ou *bfp* et/ou *EHEC-hly*A et ont été détectés à 12%, les gènes de virulence de ETEC ont été détectés sur 8% des échantillons et sont *elt* et/ou *est*Ib et/ou *est*Ia. Enfin les gènes de virulence des EAEC qui sont *pic* ou *agg*R ont été détectés dans 4% sur les échantillons. La méthodologie détaillée et les résultats de ce travail ont fait l'objet d'une publication (**Article 2**), d'une communication et d'un poster (en annexe 5).

IV. Article 2: Diarrheagenic *Escherichia coli* detected by 16-plex PCR in raw meat and beef intestines sold at local markets in Ouagadougou, Burkina Faso. [*International Journal of Food Microbiology*, 153: 154–158 (2012)].

International Journal of Food Microbiology 153 (2012) 154–158

Contents lists available at SciVerse ScienceDirect

International Journal of Food Microbiology

journal homepage: www.elsevier.com/locate/ijfoodmicro

Diarrheagenic *Escherichia coli* detected by 16-plex PCR in raw meat and beef intestines sold at local markets in Ouagadougou, Burkina Faso

Assèta Kagambèga [a,b], Outi Martikainen [a], Taru Lienemann [a], Anja Siitonen [a], Alfred S. Traoré [b], Nicolas Barro [b], Kaisa Haukka [a,*]

[a] *Bacteriology Unit, Department of Infectious Disease Surveillance and Control, National Institute for Health and Welfare (THL), P.O. Box 30, FI-00271 Helsinki, Finland*
[b] *Laboratoire de Biologie Moléculaire et d'Epidémiologie et de Surveillance Bactéries et Virus Transmis par les Aliments, CRSBAN, Département de Biochimie-Microbiologie, UFR-SVT, Université de Ouagadougou, 03 B.P. 7021 Ouagadougou 03, Burkina Faso*

ARTICLE INFO

Article history:
Received 26 May 2011
Received in revised form 20 September 2011
Accepted 24 October 2011
Available online 12 November 2011

Keywords:
E. coli pathogroups
Raw meat
16-plex PCR

ABSTRACT

The study investigated the prevalence of five major *Escherichia coli* pathogroups in raw meats and beef intestines sold at the local markets in Ouagadougou, Burkina Faso. One hundred and twenty samples (36 beef, 36 beef intestine, 24 mutton and 24 chicken samples) were purchased from four markets between October 2008 and February 2009. Fifteen virulence genes specific for Shiga toxin-producing *E. coli* (STEC), enteropathogenic *E. coli* (EPEC), enterotoxigenic *E. coli* (ETEC), enteroinvasive *E. coli* (EIEC) and enteroaggregative *E. coli* (EAEC) were examined using 16-plex PCR for mixed bacterial cultures derived from the samples. One or more diarrheagenic *E. coli* pathogroup was detected in 51 (43%) of all the 120 samples: in 16 (44%) beef, 19 (53%) beef intestine, 9 (38%) mutton and in 7 (29%) chicken samples. Thirty three (28%) samples were positive for stx_1 and/or stx_2 indicating presence of STEC. EPEC virulence markers (*eae*, *escV* and/or *ent* and/or *bfp* and/or EHEC-*hlyA*) were detected in 14 (12%) *stx*-negative samples. ETEC virulence markers (*elt* and/or *estlb* and/or *estla*) were detected in 10 (8%) samples and EAEC virulence markers (*pic* or *aggR*) in 5 (4%) samples. No EIEC was detected. The results show that in Burkina Faso the microbiological quality of retail meat is alarmingly poor due to the common occurrence of diarrheagenic *E. coli* bacteria.

© 2011 Elsevier B.V. All rights reserved.

1. Introduction

Meat and meat products have been implicated in disease outbreaks caused by various diarrheagenic *Escherichia coli* all over the world (FAO, 2005; Karmali et al., 2010; Rhoades et al., 2009). These pathogens can be classified into several pathogroups based on their virulence traits. The five most common pathogroups are: Shiga toxin-producing *E. coli* (STEC), enteropathogenic *E. coli* (EPEC), enterotoxigenic *E. coli* (ETEC), enteroinvasive *E. coli* (EIEC) and enteroaggregative *E. coli* (EAEC) (Levine, 1987; Nataro and Kaper, 1998).

STEC produces Shiga toxin encoded by stx_1 or stx_2 or their variants. Besides the *stx* gene(s), STEC strains often carry the *eae* gene, encoding the adherence factor intimin. They also often harbor additional virulence genes such as that for enterohemolysin (*EHEC-hly*) (Schmidt et al., 1995). STEC can cause gastroenteritis that may be complicated by hemorrhagic colitis or the hemolytic-uremic syndrome (HUS), which is the main cause of acute renal failure in children (Paton and Paton, 1998).

EPEC produces characteristic histopathology known as attaching and effacing (A/E) on intestinal cells (Schmidt, 2010). EPEC is further divided into two subtypes, typical (tEPEC) and atypical (aEPEC), depending on the presence or absence of the EPEC adherence factor (EAF) plasmid and *bfpB* gene (Nataro and Kaper, 1998; Schmidt, 2010). Strains of aEPEC occur most frequently in developed countries, whereas tEPEC is the leading cause of infantile diarrhea in developing countries (Trabulsi et al., 2002). ETEC produces heat-labile (LT) and/or heat-stable (ST) enterotoxin and is an important cause of diarrhea in infants and travelers (Kaper et al., 2004). EIEC is associated with invasive, bloody diarrhea resembling that caused by *Shigella* spp. Invasion is mediated by the genes located in virulence plasmid pINV encoding, e.g. Ipa proteins and their transcription regulator *invE* (Lan et al., 2004; Nataro and Kaper, 1998). EAEC harbors the mechanism for aggregative-adherence pattern mediated by aggregative adhesive fimbriae. It is increasingly recognized as a diarrheal pathogen in developing countries (Huang et al., 2004).

Domestic ruminants, mainly cattle, sheep and goats, have been established as major natural reservoirs for STEC and they play a significant role in the epidemiology of human infections (Griffin and Tauxe, 1991). During the processing of the carcasses, fecal contamination and transfer of bacteria from the animal's hide to the carcass can facilitate transmission of pathogenic *E. coli* to the meat (Elder et al., 2000). In most of the developing countries, hygienic conditions are severely

* Corresponding author at: National Institute for Health and Welfare, Department of Infectious Disease Surveillance and Control, Bacteriology Unit, P.O. Box 30, FI-00271 Helsinki, Finland. Tel.: +358 20 610 8172; fax: +358 20 610 8238.
E-mail address: kaisa.haukka@thl.fi (K. Haukka).

compromised, and living with the domestic animals within the same premises is a common practice in both rural and urban areas (Islam et al., 2010). Thus, the socioeconomic status and living style of the people in developing countries can support the occurrence of pathogenic *E. coli* infections. In Burkina Faso, pathogenic *E. coli* have been indicated to be among the main causes of diarrhea (Simporé et al., 2009). Our previous study revealed 100% prevalence of *E. coli* on raw meat vended at open markets in Ouagadougou (Kagambega et al., 2011). However, there is no data concerning the prevalence of diarrheagenic *E. coli* in food stuffs in Burkina Faso. Therefore, the objective of this study was to determine the prevalence of STEC, EPEC, ETEC, EIEC and EAEC in raw beef, beef intestine, mutton and chicken using a multiplex PCR method for detection of their virulence gene patterns.

2. Materials and methods

2.1. Samples, their processing and cultivation

The samples were purchased from four open markets in Ouagadougou, the capital city of Burkina Faso, between October 2008 and February 2009, as described previously (Kagambega et al., 2011). Briefly, 200 g of beef (n=36), beef intestine (n=36), mutton (n=24) and an entire chicken carcass (n=24) were purchased from the sellers and placed in sterile plastic bags. The samples were transported to the laboratory on ice and processed within 2 h after the collection.

In the laboratory, 25 g of each meat or intestine sample (for chicken a mixture of meat from neck, breast, wings and legs) was added into 225 ml of Buffered Peptone Water (BPW) (Liofilchem, Teramo, Italy), and homogenized using a stomacher (400 Circulator, Seward, London, UK). Ten microlitres of the mixture was streaked onto Müller Hinton (MH) agar (Liofilchem) plates and incubated at 37 °C for 24 h. The bacterial mass was collected and conserved in tubes containing trypticase soy agar at 4 °C, and subsequently sent to the National Institute for

Health and Welfare (THL) in Finland. There, the bacterial mass from the tubes was re-cultivated onto Cystine-Lactose Electrolyte-Deficient (CLED) agar plates (Difco, Sparks, USA) that were incubated at 36 °C for 18 h.

2.2. 16-plex PCR assay

The presence of STEC, EPEC, ETEC, EIEC and EAEC on meat and beef intestine samples was detected by 16-plex PCR for the genes *uidA*, *pic*, *bfp*, *invE*, *hlyA*, *elt*, *ent*, *escV*, *eaeA*, *ipaH*, *aggR*, *stx1*, *stx2*, *estIa*, *estIb* and *ast*. The primers and PCR conditions were previously described (Antikainen et al., 2009). The nucleotide sequences and predicted sizes of the amplified products for the specific oligonucleotide primers used in this study are shown in Table 1. The following criteria for identification of *E. coli* pathogroups were used: for STEC, the presence of *stx1* and/or *stx2* and possibly *eaeA*, *escV*, *ent* and *EHEC-hly*; for EPEC the presence of *eaeA* and possibly *escV*, *ent* and *bfpB* (the absence of *bfpB* indicated aEPEC); for ETEC, the presence of *elt* and/or *estIa* or *estIb*; for EIEC, the presence of *invE* and *ipaH*; for EAEC, the presence of *pic* and/or *aggR*.

For DNA extraction, a loopful of bacterial growth was taken from the first streaking area of the plate. It was suspended into 250 µl of sterile water in an Eppendorf tube, boiled at 100 °C for 10 min, and centrifuged. The supernatant was used as a PCR template. Each PCR mixture contained 2 µl of 5xGC buffer (F-519L, Finnzymes, Espoo, Finland), 0.4 µl of a mixture of dNTPs (5 mM), 2 µl of a mixture of the 16 primer pairs at the concentrations listed in Table 1, 0.24 µl of MgCl₂ (25 mM), 0.08 µl of Phusion polymerase (2 U/µl) (F-540L, Finnzymes) and 1 µl of the template. PCR-grade water was added to bring the final volume to 10 µl. The cycling conditions used in the thermal cycler (Bio-Rad Laboratories, Hercules, CA, USA) were 98 °C for 30 s, 35 cycles of 98 °C for 30 s, 63 °C for 60 s, 72 °C for 90 s with a final extension at 72 °C for 10 min. The amplified PCR products were separated by agarose gel (1.5% w/v) electrophoresis (PowerPac Basic, BioRad, CA, USA) and

Table 1
Oligonucleotide primers used for detection of the virulence genes.

Pathogroups	Target gene	Primer sequence (5′ to 3′)	Product size (bp)	Concentration (µM)	References
STEC, EPEC	eaeA	eae-F: TCAATGCAGTTCCGTTATCAGTT	482	0.1	Vidal et al. (2005)
		eae-R: GTAAAGTCCGTTACCCCAACCTG		0.1	
	escV	MP3-escV-F: ATTCTGGCTCTCTTCTTCTTTATGGCTG	544	0.4	Müller et al. (2007)
		MP3-escV-R: CGTCCCCTTTTACAAACTTCATCGC		0.4	
	ent	ent-F: TGGGCTAAAAGAAAGACACACTG	629	0.4	Müller et al. (2007)
		ent-R: CAAGCATCCTGATTATCTCACC		0.4	
Typical EPEC	bfpB	MP3-bfpB-F: GACACCTCATTGCTGAAGTCG	910	0.1	Müller et al. (2007)
		MP3-bfpB-R: CCAGAACACCTCGTTATGC		0.1	
STEC	EHEC-hly	hlyEHEC-F: TTCTGGGAAACAGTGACGCACATA	688	0.1	Antikainen et al. (2009)
		hlyEHEC-R: TCACCGATCTTCTCATCCCAATG		0.1	
	stx1	MP4-stx1A-F: CGATGTTACGGTTTGTTACTGTGACAGC	244	0.2	Müller et al. (2007)
		MP4-stx1A-R: AATGCCACGCTTCCCAGAATTG		0.2	
	stx2	MP3-stx2A-F:GTTTTGACCATCTTCGTCTGATTATTGAG	324	0.4	Müller et al. (2007)
		MP3-stx2A- R: AGCGTAAGGCTTCTGCTCTGAC		0.4	
EIEC	ipaH	ipaH-F: GAAAACCCTCCTGGTCCATCAGG	437	0.1	Antikainen et al. (2009)
		ipaH-R: GCCGGTCAGCCACCCTCTGAGAGTAC		0.1	Brandal et al. (2007)
	invE	MP2-invE-F: CGATAGATGGCGAGAAATTATATCCCG	766	0.2	Müller et al. (2007)
		MP2-invE-R: CGATCAAGAATCCCTAACAGAAGAATCAC		0.2	
EAEC	aggR	MP2-aggR-F: ACGCAGAGTTGCCTGATAAAG	400	0.2	Müller et al. (2007)
		MP2-aggR-R: AATACAGAATCGTCAGCATCAGC		0.2	
	pic	MP2-pic-F: AGCCGTTTCCGCAGAAGCC	1111	0.2	Müller et al. (2007)
		MP2-pic-R: AAATGTCAGTGAACCGACGATTGG		0.2	
	astA	MP2-astA-F: TGCCATCAACACAGTATATCCG	102	0.4	Müller et al. (2007)
		MP2-astA-R: ACGGCTTTGTAGTCCTTCCAT		0.4	
ETEC	elt	MP2-LT-F: GAACAGGAGGTTTCTGCGTTAGGTG	655	0.1	Müller et al. (2007)
		MP2-LT-R: CTTTCAATGGCTTTTTTTGGGAGTC		0.1	
	estIa	MP4-STIa-F:CCTCTTTTAGYCAGACARCTGAATCASTTG	157	0.4	Müller et al. (2007)
		MP4-STIa-R: CAGGCAGGATTACAACAAAGTTCACAG		0.4	
	estIb	MP2-STI-F: TGTCTTTTTCACCTTTCGCTC	171	0.2	Müller et al. (2007)
		MP2-STI-R: CGGTACAAGCAGGATTACAACAC		0.2	
E. coli	uidA	MP2-uidA-F: ATGCCAGTCCAGCGTTTTTGC	1487	0.2	Müller et al. (2007)
		MP2-uidA-R: AAAGTGTGGGTCAATAATCAGGAAGTG		0.2	

STEC, Shiga toxin-producing *E. coli*; EPEC, enteropathogenic *E. coli*; EIEC, enteroinvasive *E. coli*; EAEC, enteroaggregative *E. coli*; ETEC, enterotoxigenic *E. coli*.

visualized under UV light (Alpha Innotech Corporation, CA, USA) after staining with ethidium bromide.

The following reference strains were used (Table 2; Fig. 1): RHE 4283 (E 2348/69, [Baldini et al., 1983]) for EPEC, RHE 3533 (ATCC 35401) for ETEC, RHE 4270 (ATCC 43895) for STEC, RHE 6647 (145-46-215, Statens Serum Institute [SSI], Copenhagen, Denmark) for EIEC and IH 56822 (patient isolate, [Keskimäki et al., 2000]) for EAEC. The negative controls were E. coli RHE 6715 (ATCC 25922) and sterile distilled water. All the 16-plex PCR positive results were confirmed by single PCRs.

3. Results

16-plex PCR was used to detect virulence genes carried by diarrheagenic E. coli and to classify the strains as STEC, EPEC, ETEC, EIEC, or EAEC. Diarrheagenic E. coli were detected in 51 (43%) of the 120 meat samples; in 16 (44%) beef, in 19 (53%) beef intestine, in 9 (38%) mutton and in 7 (29%) chicken samples (Tables 2 and 3). Of the 33 STEC positive samples, 12 (36%) contained stx1, stx2, eae and EHEC-hlyA, 11 (32%) stx1 and stx2, 8 (22%) only stx1, and 2 (6%) only stx2. One typical EPEC containing eae, escV, ent and bfp was found in chicken. All the other EPECs were atypical and possessed eae, escV and/or ent. Two EPECs from mutton samples contained also EHEC-hlyA. One mutton sample was negative for eae but positive for escV and ent and thus interpreted as containing EPEC. The detected ETEC strains contained estIa, estIb and/or elt. The five detected EAEC strains contained aggR but no pic.

In 23 samples (19%) only STEC was detected, in 13 (11%) only EPEC, in 3 (3%) only ETEC and in 11% only EAEC (Table 3). No EIEC virulence determinants were detected in any of the samples. STEC only was detected in 25% of the beef, in 27% of the beef intestine, in 16% of the mutton and in 0% of the chicken samples (Table 3). Correspondingly, EPEC only was detected in 11%, 0%, 13% and 25%; ETEC only in 0%, 8%, 0%, and 0%; EAEC only in 0%, 0%, 4% and 0% of the samples. In several samples the virulence determinants for two pathogroups were observed (Table 3); STEC and ETEC were detected in 6 (5%) samples, STEC and EAEC in 4 (3%) and aEPEC and ETEC in 1 (1%). Determination of the co-presence of STEC and EPEC was not possible because STEC strains may contain all the virulence genes present in atypical EPEC.

4. Discussion

The microbiological safety of meat and meat products is an important public health concern worldwide. In the urban areas of Burkina Faso, meat and meat products are consumed daily, although their microbiological quality is questionable (Kagambega et al., 2011).

Based on an earlier survey in Burkina Faso, Barro et al. (2005) suggested that the consumption of meat products is among the major causes of diarrheal diseases. In general, the microbiological safety of food stuffs sold at common markets in developing countries has rarely been studied. The present study was the first conducted in Burkina Faso to detect diarrheagenic E. coli based on the presence of the virulence genes in bacterial cultures derived from foodstuffs. Although for a proper risk assessment the studied samples are too few, our results showed that the microbiological quality of retail meat in Burkina Faso is alarmingly poor due to the common occurrence of enteropathogenic E. coli bacteria.

Of the five E. coli pathogroups investigated, STEC was detected at the highest rate, 34% of the beef, 43% of the beef intestine and 20% of the mutton samples were STEC-positive. The results suggest that STEC is more prevalent in meat sold in Burkina Faso than in most of the other countries. In general, prevalence of STEC has been found to decline throughout the abattoir process as the hides are removed and antimicrobial interventions applied, so that the final prevalence on chilled carcass sides is around 0.4% (Rhoades et al., 2009). In Burkina Faso chilling of carcasses is not used at any stage of the meat processing or retail, which might explain the high prevalence of STEC detected in the study. Also the method used can affect the reported percentages. Other surveys using direct PCR-detection have reported such prevalences as 11% for beef in France (Pradel et al., 2000) and 0.23% for the retail meat in the US (Xia et al., 2010). In Nigeria, the prevalence of the selected 7 STEC serogroups in meat samples was up to 4% (Ojo et al., 2010). Also in other studies conducted in less developed countries, lower prevalences than ours have been obtained when PCR detection of the various pathogroups has been used on E. coli isolates (Lee et al., 2009; Rúgeles et al., 2010). Worldwide, infections caused by STEC have often been linked especially to the consumption of ground beef (Hussein, 2007). Its low infectious dose compared to EPEC, ETEC and EAEC emphasizes the importance of its surveillance in retail foods.

EPEC, ETEC and EAEC were detected less often than STEC in our meat and intestine samples, except in chickens, which seem to be the major carriers of atypical EPEC. Similar results have been obtained in the studies conducted in Cameroon (Nzouankeu et al., 2010) and in India (Farooq et al., 2009). Interestingly, atypical EPEC was detected in 11% of our beef samples, while no EPEC positive beef intestine samples were detected. One explanation to this discrepancy might be that by using the 16-plex PCR detection it was not possible to determine the co-presence of STEC and atypical EPEC, because STEC strains can contain all the virulence genes present in atypical EPEC. Among the intestine samples, eaeA and escV indicative of EPEC were present in 8% of the samples, which contained also stx indicative of STEC. The samples

Table 2
Virulence genes detected by 16-plex PCR in 51 of the 120 studied meat and intestine samples and in the six control strains.

Pathogroups	Virulence genes															
	eae	escV	ent	bfp	EHEC-hly	stx1	stx2	ipaH	invE	aggR	pic	astA	elt	estIa	estIb	uidA
	Control strains															
STEC	RH4270	+	+	+	−	+	−	+	+	−	−	−	−	−	−	−	+
EPEC	RH428	+	+	+	−	−	−	−	−	−	−	−	−	−	−	−	−
ETEC	RH3533	−	−	−	−	−	−	−	−	−	−	−	+	+	−	+	+
EAEC	IH56822	−	−	−	−	−	−	−	−	+	+	+	+	−	−	−	+
EIEC	RH6647	−	−	−	−	−	+	+	−	−	−	−	−	−	−	−	+
Neg. control	RHE6715	−	−	−	−	−	−	−	−	−	−	−	−	−	−	−	+
	Number of positive samples															
STEC	23	6	7	3	−	15	18	19	−	−	−	−	20	−	−	−	23
tEPEC	1	1	1	1	1	−	−	−	−	−	−	−	1	−	−	−	1
aEPEC	12	11	12	7	−	2	−	−	−	−	−	−	12	−	−	−	9
ETEC	3	−	−	−	−	−	−	−	−	−	−	−	3	2	2	−	3
EAEC	1	−	−	−	−	−	−	−	−	−	1	−	1	−	−	−	1
STEC + ETEC	6	1	1	−	−	6	6	1	−	−	−	−	6	4	2	−	4
STEC + EAEC	4	−	2	1	−	3	4	2	−	−	4	−	4	−	−	−	4
aEPEC + ETEC	1	1	1	−	−	−	−	−	−	−	−	−	−	−	−	1	−

tEPEC, typical EPEC; aEPEC, atypical EPEC.

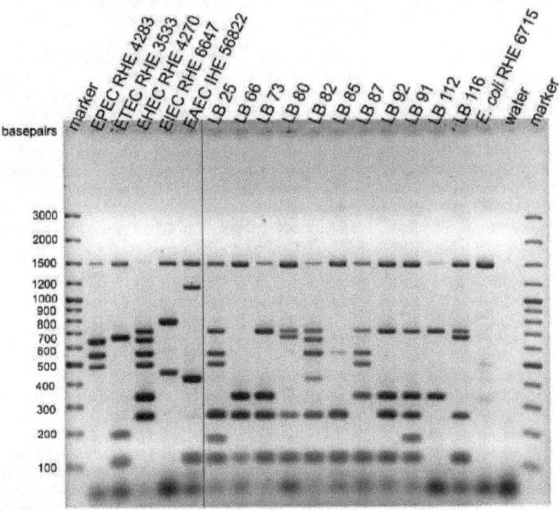

Fig. 1. Example of the 16-plex PCR results. LB25, beef intestine sample containing STEC + ETEC; LB66, beef STEC; LB73, beef STEC; LB80, beef STEC + ETEC; LB82, beef STEC + EAEC; LB85, beef STEC; LB87, beef STEC; LB92, beef STEC; LB91, beef STEC + ETEC; LB112, mutton STEC; LB116, mutton STEC + ETEC.

were consequently interpreted as STEC positive but it is possible that they also contained atypical EPEC (as is also the case with two STEC-positive beef samples). Indeed, in the previous studies most STEC isolates from healthy cattle have been found not to carry the *eae* gene (Xia et al., 2010).

Only one typical EPEC was detected among our meat samples (1%), whereas atypical EPEC was detected in 13 samples (11%). Typical EPEC is known to occur rarely in animals (Nataro and Kaper, 1998). Also in the feces of diarrheal children in Burkina Faso atypical EPEC was more common (12%) than typical EPEC (4%) (Bonkoungou et al., in press). Atypical EPEC appear to be more closely related to STEC and they are both considered emerging pathogens (Beutin et al., 2003; Trabulsi et al., 2002). Among our mutton samples, we had two which were EHEC-*hly*-positive but *stx*-negative, one of which was also *eae*-positive. *eae*-positive and *stx*-negative atypical EPEC strains that possess EHEC-*hly* have been reported from cattle and sheep also before (Cookson et al., 2007).

In the virulence of STEC, *stx2* is considered to be the most important virulence factor associated with HUS (Boerlin et al., 1999; Ostroff et al., 1989). In our study, the proportion of STEC carrying both *stx1* and *stx2* was 32%; only *stx1* was carried by 22% and only *stx2* by 6% of STEC. In comparison, Lee et al. (2009) observed that 64% of the STEC from raw meat carried *stx2*, 14% carried *stx1* and *stx2*, 14% carried *stx2* and *eaeA*, and 7% carried *stx1* and *eaeA*.

In the present study, virulence genes for more than one DEC pathogroup were detected in many samples. Most probably this is due to contamination of the meat by more than one pathogroup. However, it is also possible that in some cases the meat contains a so-called hybrid strain that has, as a result of horizontal gene transfer, gained virulence genes from another DEC pathogroup. Recently, an unusual gene combination of enterohemorrhagic E. coli O104:H4 was found to be a cause of the large and severe outbreak in Germany (EFSA, 2011). The strain involved had the chromosomal backbone similar to EAEC, but it also carried a *stx2* gene associated with STEC and HUS (Mellmann et al., 2011).

Table 3
Prevalence of diarrheagenic E. coli (%) in 120 meat and intestine samples.

E. coli pathogroups	Samples tested				Total n = 120
	Beef n = 36	Beef intestine n = 36	Mutton n = 24	Chicken n = 24	
Any DEC	16 (44%)	19 (53%)	9 (38%)	7 (29%)	51 (43%)
STEC only	9 (25%)	10 (27%)	4 (16%)	0	23 (19%)
EPEC only	4 (11%)	0	3 (13%)	6 (25%)	13 (11%)
ETEC only	0	3 (8%)	0	0	3 (3%)
EAEC only	0	0	1 (4%)	0	1 (1%)
STEC + ETEC	2 (6%)	3 (8%)	1 (4%)	0	6 (5%)
STEC + EAEC	1 (3%)	3 (8%)	0	0	4 (3%)
EPEC + ETEC	0	0	0	1 (4%)	1 (1%)

DEC, Diarrheagenic Escherichia coli.

Thus, E. coli strains are capable in developing new gene combinations that can lead into the emergence of more virulent clones. It is important to detect other pathogroups besides STEC and findings that contain both STEC and EAEC virulence genes, as detected in four samples of our study, warrant further investigation.

Our work demonstrates that raw meat sold at open markets in Ouagadougou commonly contains diarrheagenic E. coli pathogroups. The presence of these pathogens can be due to contamination taking place during the meat processing at slaughterhouse or to the retailers' poor handling of meat. Consequently, it is necessary to effectively prevent contamination by educating employees, retailers and consumers on the appropriate handling, storage and heating of meat. Simple intervention strategies, such as promoting hand washing with soap and good hygienic practices at the slaughterhouses, processing plants and retailer shops, can have a sound practical impact on public health.

Acknowledgments

The study was funded by the Academy of Finland grant 122600 to collaboration between the Finnish National Institute for Health and Welfare (THL) and CRSBAN/University of Ouagadougou, Burkina Faso.

References

Antikainen, J., Tarkka, E., Haukka, K., Siitonen, A., Vaara, M., Kirveskari, J., 2009. New 16-plex PCR method for rapid detection of diarrheagenic Escherichia coli directly from stool samples. European Journal of Clinical Microbiology and Infectious Diseases 28, 899–908.

Baldini, M.M., Kaper, J.B., Levine, M.M., Candy, D.C., Moon, H.W., 1983. Plasmid-mediated adhesion in enteropathogenic Escherichia coli. Journal of Pediatric Gastroenterology and Nutrition 2, 534–538.

Barro, N., Sangaré, L., Tahita, M., Ouattara, C.A.T., Traoré, A.S., 2005. Les principaux agents du péril fécal identifiés dans les aliments de rue et ceux des cantines Burkina Faso et ailleurs et les risques de maladies associées. Colloque Régional scientifique et Pédagogique: Maîtrise de procédés en vue d'améliorer la qualité et la sécurité des aliments, Utilisation des OGM, analyses des risques en Agroalimentaires UO/AUF/GP3A/CIDEFA: Du 8 au 10 novembre 2005 à Ouagadougou, Burkina Faso.

Beutin, L., Marchés, O., Bettelheim, K.A., Gleier, K., Zimmermann, S., Schmidt, H., Oswald, E., 2003. HEp-2 cell adherence, actin aggregation, and intimin types of attaching and effacing Escherichia coli strains isolated from healthy infants in Germany and Australia. Infection and Immunity 71, 3995–4002.

Boerlin, P., McEwen, S.A., Boerlin-Petzold, F., Wilson, J.B., Johnson, R.P., Gyles, C.L., 1999. Association between virulence factors of Shiga toxin-producing Escherichia coli and disease in humans. Journal of Clinical Microbiology 37, 497–503.

Bonkoungou, I.J.O., Lienemann, T., Martikainen, O., Dembele, R., Sanou, I., Traore, A., Siitonen, A., Barro, N., Haukka, K., in press. Diarrhoeagenic Escherichia coli detected by 16-plex PCR in children with and without diarrhoea in Burkina Faso. Clinical Microbiology and Infection(Epub ahead of print). doi:10.1111/j.1469-0691.2011.03675.x

Brandal, L.T., Lindstedt, B.A., Aas, L., Stavnes, T.L., Lassen, J., Kapperud, G., 2007. Octaplex PCR and fluorescence-based capillary electrophoresis for identification of human diarrheagenic Escherichia coli and Shigella spp. Journal of Microbiological Methods 68, 331–341.

Cookson, A.L., Bennett, J., Thomson-Carter, F., Attwood, G.T., 2007. Molecular subtyping and genetic analysis of the enterohemolysin gene (ehxA) from Shiga toxin-producing Escherichia coli and atypical enteropathogenic E. coli. Applied and Environmental Microbiology 73, 6360–6369.

Elder, R.O., Keen, J.E., Siragusa, G.R., Barkocy-Gallagher, G.A., Koohmaraie, M., Laegreid, W. W., 2000. Correlation of enterohemorrhagic Escherichia coli O157 prevalence in feces, hides, and carcasses of beef cattle during processing. Proceedings of the National Academy of Sciences of the United States of America 97, 2999–3003.

European Food Safety Authority (EFSA), 2011. Urgent advice on the public health risk of Shiga-toxin producing Escherichia coli in fresh vegetables. EFSA Journal 9, 2274.

Farooq, S., Hussain, I., Mir, M.A., Bhat, M.A., Wani, S.A., 2009. Isolation of atypical enteropathogenic Escherichia coli and Shiga toxin 1 and 2f-producing Escherichia coli from avian species in India. Letters in Applied Microbiology 48, 692–697.

Food and Agriculture Organization (FAO), 2005. Code of Hygienic Practice for Meat CAC/RCP 58-200.

Griffin, P.M., Tauxe, R.V., 1991. The epidemiology of infections caused by Escherichia coli O157:H7, other enterohemorrhagic E. coli, and the associated hemolytic uremic syndrome. Epidemiologic Reviews 13, 60–97.

Huang, D.B., Okhuysen, P.C., Jiang, Z., Dupont, H.L., 2004. Enteroaggregative Escherichia coli: an emerging enteric pathogen. The American Journal of Gastroenterology 99, 383–389.

Hussein, H.S. 2007. Prevalence and pathogenicity of Shiga toxin-producing Escherichia coli in beef cattle and their products. Journal of Animal Science 85, E63–E72.

Islam, M.A., Mondol, A.S., Azmi, I.J., de Boer, E., Beumer, R.R., Zwietering, M.H., Heuvelink, A.E., Talukder, K.A., 2010. Occurrence and characterization of Shiga toxin-producing Escherichia coli in raw meat, raw milk, and street vended juices in Bangladesh. Foodborne Pathogens and Disease 7, 1381–1385.

Kagambega, A., Haukka, K., Siitonen, A., Traore, A., Barro, N., 2011. Prevalence of Salmonella enterica and the hygienic indicator Escherichia coli in raw meat at markets in Ouagadougou, Burkina Faso. Journal of Food Protection 74, 1547–1551.

Kaper, J.B., Nataro, J.P., Mobley, H.L., 2004. Pathogenic Escherichia coli. Nature Reviews. Microbiology 2, 123–140.

Karmali, M.A., Gannon, V., Sargeant, J.M., 2010. Verocytotoxin-producing Escherichia coli (VTEC). Veterinary Microbiology 140, 360–370.

Keskimäki, M., Mattila, L., Peltola, H., Siitonen, A., 2000. Prevalence of diarrheagenic Escherichia coli in Finns with or without diarrhea during a round-the-world trip. Journal of Clinical Microbiology 38, 4425–4429.

Lan, R., Alles, M.A., Donohoe, K., Martinez, M.B., Reeves, P.R., 2004. Molecular evolutionary relationships of enteroinvasive Escherichia coli and Shigella spp. Infection and Immunity 72, 5080–5088.

Lee, G.Y., Jang, H.I., Hwang, I.G., Rhee, M.S., 2009. Prevalence and classification of pathogenic Escherichia coli isolated from fresh beef, poultry and pork in Korea. International Journal of Food Microbiology 134, 196–200.

Levine, M.M., 1987. Escherichia coli that cause diarrhea: enterotoxigenic, enteropathogenic, enteroinvasive, enterohemorrhagic, and enteroadherent. The Journal of Infectious Diseases 155, 377–389.

Mellmann, A., Harmsen, D., Cummings, C.A., Zentz, E.B., Leopold, S.R., Rico, A., Prior, K., Szczepanowski, R., Ji Y., Zhang, W., McLaughlin, S.F., Henkhaus, J.K., Leopold, B., Bielaszewska, M., Prager, R., Brzoska, P.M., Moore, R.L., Guenther, S., Rothberg, J.M., Karch, H., 2011. Prospective genomic characterization of the German enterohemorrhagic Escherichia coli O104:H4 outbreak by rapid next generation sequencing technology. PLoS One 6, e22751.

Müller, D., Greune, L., Heusipp, G., Karch, H., Fruth, A., Tschäpe, H., Schmidt, H.A., 2007. Identification of unconventional intestinal pathogenic Escherichia coli isolates expressing intermediate virulence factor profiles by using a novel single-step multiplex PCR. Applied and Environmental Microbiology 73, 3380–3390.

Nataro, J.P., Kaper, J.B., 1998. Diarrheagenic Escherichia coli. Clinical Microbiology Reviews 11, 142–201.

Nzouankeu, A., Ngandjio, A., Ejenguele, G., Njine, T., Ndayo Wouafo, M., 2010. Multiple contaminations of chickens with Campylobacter, Escherichia coli and Salmonella in Yaounde (Cameroon). Journal of Infection in Developing Countries 4, 583–686.

Ojo, O.E., Ajuwape, A.T.P., Otesile, E.B., Owoade, A.A., Oyekunle, M.A., Adetosoye, A.I., 2010. Potentially zoonotic shiga toxin-producing Escherichia coli serogroups in the faeces and meat of food-producing animals in Ibadan, Nigeria. International Journal of Food Microbiology 142, 214–221.

Ostroff, S., Tarr, P., Neill, M., Lewis, J., Hargett-Bean, N., Kobayashi, J., 1989. Toxin genotypes and plasmid profiles as determinants of systemic sequelae in Escherichia coli O157:H7 infections. The Journal of Infectious Diseases 160, 994–999.

Paton, J.C., Paton, A.W., 1998. Pathogenesis and diagnosis of Shiga toxin-producing Escherichia coli infections. Clinical Microbiology Reviews 11, 450–479.

Pradel, N., Livrelli, V., De Champs, C., Palcoux, J.B., Reynaud, A., Scheutz, F., Sirot, J., Joly, B., Forestier, C., 2000. Prevalence and characterization of Shiga toxin-producing Escherichia coli isolated from cattle, food, and children during a one-year prospective study in France. Journal of Clinical Microbiology 38, 1023–1031.

Rhoades, J.R., Duffy, G., Koutsoumanis, K., 2009. Prevalence and concentration of verocytotoxigenic Escherichia coli, Salmonella enterica and Listeria monocytogenes in the beef production chain: a review. Food Microbiology 26, 357–376.

Rúgeles, L.C., Bai, J., Martínez, A.J., Vanegas, M.C., Gómez-Duarte, O.G., 2010. Molecular characterization of diarrheagenic Escherichia coli strains from stools samples and food products in Colombia. International Journal of Food Microbiology 138, 282–286.

Schmidt, M.A., 2010. LEE ways: tales of EPEC, ATEC and EHEC. Cellular Microbiology 12, 1544–1552.

Schmidt, H., Beutin, L., Karch, H., 1995. Molecular analysis of the plasmid-encoded hemolysin of Escherichia coli O157:H7 strain EDL 933. Infection and Immunity 63, 1055–1061.

Simporé, J., Ouermi, D., Ilboudo, D., Kabre, A., Zeba, B., Pietra, V., Pignatelli, S., Nikiema, J. B., Kabre, G.B., Caligaris, S., Schumacher, F., Castelli, F., 2009. Aetiology of acute gastro-enteritis in children at Saint Camille Medical Centre, Ouagadougou, Burkina Faso. Pakistan Journal of Biological Sciences 12, 258–263.

Trabulsi, L.R., Keller, R., Tardelli Gomes, Tardelli Gomes, T.A., 2002. Typical and atypical enteropathogenic Escherichia coli. Emerging Infectious Diseases 8, 508–513.

Vidal, M., Kruger, E., Durán, C., Lagos, R., Levine, M., Prado, V., Toro, C., Vidal, R., 2005. Single multiplex PCR assay to identify simultaneously the six categories of diarrheagenic Escherichia coli associated with enteric infections. Journal of Clinical Microbiology 43, 5362–5365.

Xia, X., Meng, J., McDermott, P.F., Ayers, S., Blickenstaff, K., Tran, T.T., Abbott, J., Zheng, J., Zhao, S., 2010. Presence and characterization of shiga toxin-producing Escherichia coli and other potentially diarrheagenic E. coli strains in retail meats. Applied and Environmental Microbiology 76, 1709–1717.

V. Contamination des carcasses de poulet par *Salmonella* et *E. coli*

Il est ressorti au cours des premières investigations que les poulets ont été les plus contaminés par les *Salmonella*, ce qui nous a amenés à faire une seconde investigation uniquement sur les carcasses de poulet et en plus grand nombre vue que les poulets deviennent de plus en plus utilisés par les grilleurs à travers la ville de Ouagadougou et beaucoup appréciés par les burkinabè. Pourtant, le manque de contrôle vétérinaire et d'abattoir pour la volaille a été signalé lors de l'enquête menée au cours de la première investigation.

Dans ce présent travail, la prévalence de *Salmonella* a été de 57% sur les carcasses de poulet dont les sérotypes identifiés ont été *S.* Derby, Chester, Hato, Agona, Banana, Monschui, Senftenberg, Adelaide, Anatum, Brancaster, Eastbourne, Galiema, Nima, Nottingham, Saarbruecken et Typhi. Certains de ces sérotypes ont présenté une résistance vis-à-vis des antibiotiques testés.

Les pathovars de *E. coli* ont été détectés à un taux de 45% sur les carcasses de poulet par l'utilisation de la 16-plex PCR. Les pathovars identifiés ont été EPEC (39%), STEC (6%) et EAEC (13%). Le pathovar EAEC a été seulement détecté en combinaison avec EPEC ou STEC. Les autres pathovars ETEC ou EIEC n'ont pas été détectés. Les résultats de ce travail ont fait l'objet d'un articlepublié (**Article 3**).

VI. Article 3: Characterization of *Salmonella enterica* and detection of the virulence genes specific to diarrheagenic *Escherichia coli* from poultry carcasses in Ouagadougou, Burkina Faso (*Foodborne Pathogens and Diseases, 9(7):589-93 . doi: 10.1089/fpd.2011.1071*).

FOODBORNE PATHOGENS AND DISEASE
Volume 9, Number 00, 2012
© Mary Ann Liebert, Inc.
DOI: 10.1089/fpd.2011.1071

Original Article

Characterization of *Salmonella enterica* and Detection of the Virulence Genes Specific to Diarrheagenic *Escherichia coli* from Poultry Carcasses in Ouagadougou, Burkina Faso

Assèta Kagambèga,[1,2] Nicolas Barro,[2] Alfred S. Traoré,[2] Anja Siitonen,[1] and Kaisa Haukka[1]

Abstract

One hundred chicken carcasses purchased from three markets selling poultry in Ouagadougou, Burkina Faso, between June 2010 and October 2010 were examined for their microbiological quality. The presence of *Salmonella* was investigated using standard bacteriological procedures, and the isolates obtained were serotyped and tested for antimicrobial susceptibility. The presence of virulence-associated genes of the five main pathogroups of diarrheagenic *Escherichia coli*—Shiga toxin–producing *E. coli* (STEC), enteropathogenic *E. coli* (EPEC), enteroaggregative *E. coli* (EAEC), enterotoxigenic *E. coli*, and enteroinvasive *E. coli*—was investigated using 16-plex polymerase chain reaction (PCR) on the mixed bacterial cultures from the poultry samples. Of the 100 chicken carcasses studied, 57 were contaminated by *Salmonella*; 16 different serotypes were identified, the most frequent being *Salmonella* Derby, found in 28 samples. Four *Salmonella* strains were resistant to tetracycline, and two were resistant to streptomycin. Based on the PCR detection of the virulence genes, in total, 45 carcasses were contaminated by three pathogroups of *E. coli*: STEC, EPEC, or EAEC. The STEC and EPEC virulence genes were detected on six and 39 carcasses, respectively. EAEC virulence genes were only detected in combination with those of EPEC (on 11 carcasses) or STEC (on two carcasses). The STEC-positive carcasses contained the genes stx_1, stx_2, *eaeA*, *esc*V, and *ent* in different combinations. None of the EPEC-positive carcasses contained the *bfp* gene, indicating that only atypical EPEC was present. EAEC virulence genes detected were *aggR* and/or *pic*. The high proportion of chicken carcasses contaminated by *Salmonella* and diarrheagenic *E. coli* indicates a potential food safety risk for consumers and highlights the necessity of public awareness of these pathogens.

Introduction

CHICKENS ARE WIDELY CONSUMED in Burkina Faso as well as exported to the neighboring countries. Contaminated raw or undercooked poultry products, especially chicken meat, have been shown to be an important source of foodborne pathogens for humans, resulting in numerous cases of enteric infections (Wilson, 2002). Most of the foodborne illnesses are caused by three major bacteria: *Campylobacter* spp., *Salmonella enterica*, and pathogenic *Escherichia coli* (Todd, 1997). The antimicrobial resistance of the enteric pathogens is also an important public health problem, as it is steadily increasing and can be associated with extended hospitalizations (Sackey *et al.*, 2000;

Varma *et al.*, 2005). Multiresistant bacteria such as *Salmonella* found in humans may be of animal origin, from which they are transmitted to humans through food (White *et al.*, 2001).

Diarrheagenic *E. coli* strains are commonly classified into five main heterogeneous groups based on their virulence traits (Nataro and Kaper, 1998): enterohemorrhagic *E. coli* (EHEC), also called Shiga toxin–producing *E. coli* (STEC); enteropathogenic *E. coli* (EPEC); enteroaggregative *E. coli* (EAEC); enterotoxigenic *E. coli* (ETEC); and enteroinvasive *E. coli* (EIEC). All the pathogroups can cause diarrhea and other symptoms, but especially EHEC can also cause serious complications, such as bloody diarrhea and hemolytic uremic syndrome (Nataro and Kaper, 1998).

[1]Bacteriology Unit, Department of Infectious Disease Surveillance and Control, National Institute for Health and Welfare, Helsinki, Finland.
[2]Laboratory of Molecular Biology, Epidemiology, and Bacteria and Virus Transmitted by Food, Research Centre of Biological and Food Sciences (CRSBAN)/Department of Biochemical-Microbiology, Research and Formation Unit of Live Sciences (UFR-SVT)/University of Ouagadougou, Ouagadougou, Burkina Faso.

1

Traditionally, in Burkina Faso chickens roam free, scattering their feces anywhere on the house yards. More recently, also a new business activity of small-scale industrial broiler poultry production has expanded to supply the growing urban population with its demand of animal proteins. However, the hygienic conditions at the poultry markets, where slaughtering is also done, are inadequate and can lead to bacterial contamination of meat (Kagambega et al., 2011). The objective of this study was to determine the prevalence of *Salmonella* and diarrheagenic *E. coli* in chicken carcasses purchased from retail markets in Ouagadougou during the rainy season as well as to serotype and characterize the antimicrobial resistance of the *Salmonella* isolates.

Materials and Methods

Sampling plan

Chicken carcasses were purchased from three retail markets in Ouagadougou. Ten sellers from three markets were each visited 10 times from June to October 2010, the time period that covers the rainy season. The entire carcasses slaughtered and plucked during the visit were placed in sterile plastic bags and transported in a cool box to the laboratory, where samples were processed within an hour. There were no records available concerning the origin of the chickens. But, according to the sellers, chickens were of local breeds obtained from different areas across the country.

Microbial analyses

Salmonella. The whole carcass was placed in a large plastic bag containing 225 mL of buffered peptone water, and the bag was vigorously massaged and shaken for 1 min at room temperature. Fifty milliliters of the rinse fluid from the bag was incubated at 37°C for 18–20 h, and then 0.1 mL was added to 10 mL of Rappaport-Vassiliadis broth (Oxoid, Basingstoke, UK) and incubated for an additional 24 h at 42°C before a loopful (10 μL) was plated on xylose-lysine-deoxycholate agar (Oxoid). Colonies exhibiting typical *Salmonella* morphology were preliminarily confirmed biochemically using lysine and triple sugar iron agars. Final verification was done at the National Institute for Health and Welfare (Helsinki, Finland) with API 20E (Biomerieux, Marcy l'Etoile, France), and the strains were serotyped according to the Kauffman-White scheme (Kauffmann, 1971).

All *Salmonella* strains were tested for susceptibility to 12 different antimicrobial agents using the disk diffusion method on Mueller-Hinton agar (Oxoid) at 37°C for 24 h. The antibiotic disks (Oxoid) used were ampicillin (10 μg), chloramphenicol (30 μg), streptomycin (10 μg), sulfonamide (300 μg), trimethoprim (5 μg), ciprofloxacin (5 μg), tetracycline (30 μg), gentamicin (10 μg), nalidixic acid (30 μg), cefotaxime (5 μg), mecillinam (10 μg), and imipenem (10 μg).

Diarrheagenic *E. coli*. Fifty milliliters of the buffered peptone water rinse fluid was incubated at 37°C for 18–20 h. A loopful (10 μL) of the enriched sample was streaked onto sorbitol MacConkey's agar (Oxoid) and incubated at 37°C overnight. All the bacterial mass and colonies growing on a plate were collected, conserved in 1.8-mL tubes containing trypticase soy agar at 4°C, and subsequently sent to the National Institute for Health and Welfare. There, a plastic stick

was used to retrieve some bacterial mass from the tubes, and bacteria were recultivated on cystine–lactose electrolyte-deficient agar (Difco, Sparks, MD) at 36°C for 18 h. A loopful (10 μL) of the bacterial mass from the cysteine-lactose electrolyte-deficient plate was boiled in 300 μL of sterile water for 10 min, and the supernatant containing DNA was used for polymerase chain reaction (PCR) detection of the virulence genes of diarrheagenic *E. coli*.

The presence of the specific virulence genes for STEC, EPEC, EAEC, ETEC, and EIEC in the chicken carcasses was studied using 16-plex PCR targeting the genes *uid*A, *stx₁*, *stx₂*, *hly*A, *eae*A, *esc*V, *ent*, *bfp*, *agg*R, *pic*, *elt*, *estl*a, *estl*b, *inv*E, *ipa*H, and *ast*A. The primers, PCR conditions, and control strains used were previously described by Kagambega et al. (2012). The following criteria were used for identification of *E. coli* pathogroups: for STEC, the presence of *stx₁* and/or *stx₂* and possibly *eae*A, *esc*V, *ent*, and EHEC-*hly*; for EPEC, the presence of *eae*A and possibly *esc*V, *ent*, and *bfp* (the absence of *bfp* indicated atypical EPEC); for EAEC, the presence of *pic* and/or *agg*R; for ETEC, the presence of *elt* and/or *estl*a or *estl*b; and for EIEC, the presence of *inv*E and *ipa*H. Because *ast*A was not specific for a certain pathogroup, it was not used in the final analysis.

Results

Of the 100 chicken carcasses examined, 57 were contaminated by *S. enterica* and, based on the presence of the virulence genes, 45 by diarrheagenic *E. coli*. Sixteen different serotypes of *Salmonella* were identified—Derby (28 isolates), Chester (5), Hato (4), Banana, Monschui, Senftenberg (3 of each), and Adelaide, Agona, Anatum, Brancaster, Eastbourne, Galiema, Nima, Nottingham, Saarbruecken, and Typhi (1 of each)—along with one strain of Group B (4,12:e,h:-) and one of Group C (6,7,14:d:-). Most of the isolates were sensitive to the tested antimicrobials; only four Derby isolates were resistant to tetracycline, and one Derby isolate and one Anatum isolate were resistant to streptomycin. Among the 45 samples containing the virulence genes of diarrheagenic *E. coli*, EPEC genes were detected in 28 samples, STEC genes were detected in 4 samples, EPEC genes together with EAEC genes were detected in 11 samples, and STEC genes together with EAEC genes were detected in 2 samples (Table 1). No ETEC or EIEC genes were detected. The STEC-positive carcasses contained the genes *stx₁*, *stx₂*, *eae*A, *esc*V, and *ent* in different combinations. None of the EPEC-positive carcasses harbored the *bfp* gene, indicating that only atypical EPEC was present. In addition, they carried *esc*V and/or *eae* and/or *ent* as virulence markers. The detected EAEC virulence genes were *agg*R and/or *pic*.

Discussion

Our study revealed a common occurrence of *Salmonella* (57%) and the virulence genes of diarrheagenic *E. coli* (45%) in chicken carcasses sold at the retail markets in Ouagadougou. The proportion of *Salmonella* in this study was comparable to that observed in chickens in Cameroon (60%) (Nzouankeu et al., 2010) and in Ethiopia (68%) (Tibaijuka et al., 2003). The high prevalence of *Salmonella* in chicken may be due to asymptomatic carriage of *Salmonella* in avian caeca, which can lead into cross-contamination of the carcass during or after slaughter (Tibaijuka et al., 2003; Dione et al., 2011). This is especially true when considering the poor hygienic conditions

TABLE 1. VIRULENCE GENES ASSOCIATED WITH THE PATHOGROUPS OF DIARRHEAGENIC *ESCHERICHIA COLI* DETECTED BY 16-PLEX POLYMERASE CHAIN REACTION IN CHICKEN CARCASSES

Control strains	Virulence genes														
	stx_1	stx_2	EHEC-hly	eaeA	escV	ent	bfp	elt	estIa	estIb	aggR	pic	ipaH	invE	uidA
RH4270 (STEC)	+	+	+	+	+	+	–	–	–	–	–	–	–	–	+
RH4283 (EPEC)	–	–	–	+	+	+	–	–	–	–	–	–	–	–	+
IH56822 (EAEC)	–	–	–	–	–	–	–	–	–	–	+	+	–	–	+
RH3533 (ETEC)	–	–	–	–	–	–	–	+	+	+	–	–	–	–	+
RH6647 (EIEC)	–	–	–	–	–	–	–	–	–	–	–	–	+	+	+
Identified pathogroups (n^a)															
STEC (2)	+	–	–	–	+	–	–	–	–	–	–	–	–	–	+
STEC (1)	–	+	–	–	+	–	–	–	–	–	–	–	–	–	+
STEC (1)	+	–	–	+	+	–	–	–	–	–	–	–	–	–	+
STEC+EAEC (1)	+	–	–	–	+	–	–	–	–	–	–	+	–	–	+
STEC+EAEC (1)	–	+	–	+	+	+	–	–	–	–	+	–	–	–	+
aEPEC (16)	–	–	–	–	+	–	–	–	–	–	–	–	–	–	+
aEPEC (1)	–	–	+	–	+	–	–	–	–	–	–	–	–	–	+
aEPEC (7)	–	–	–	–	+	+	–	–	–	–	–	–	–	–	+
aEPEC (4)	–	–	–	+	+	–	–	–	–	–	–	–	–	–	+
aEPEC+EAEC (4)	–	–	–	–	+	–	–	–	–	–	–	+	–	–	+
aEPEC+EAEC (1)	–	–	–	+	+	–	–	–	–	–	–	+	–	–	+
aEPEC+EAEC (1)	–	–	–	+	+	–	–	–	–	–	+	–	–	–	+
aEPEC+EAEC (3)	–	–	–	–	+	–	–	–	–	–	+	–	–	–	+
aEPEC+EAEC (1)	–	–	–	–	+	+	–	–	–	–	+	–	–	–	–

Positive (+) and negative (–) polymerase chain reaction findings are indicated.
[a]Number of the samples with the indicated virulence gene profile.
aEPEC = atypical enteropathogenic *E. coli*; EAEC, enteroaggregative *E. coli*; EIEC, enteroinvasive *E. coli*; EPEC, enteropathogenic *E. coli*; ETEC, enterotoxigenic *E. coli*; STEC, Shiga toxin–producing *E. coli* (also called enterohemorrhagic *E. coli*).

at the popular open markets where the slaughter of chickens often takes place (Kagambega et al., 2011). The study conducted in Accra, Ghana, on chicken carcasses purchased from supermarkets detected a lower proportion (7%) of *Salmonella* (Sackey et al., 2000).

In the present study, *Salmonella* Derby was the predominant serotype, as it was also in our previous study (Kagambega et al., 2011). Other serotypes that we found previously during the dry season (that is, *Salmonella* Agona and *Salmonella* Tilene) were not common this time. In other Western African countries several different serotypes have been found to be most common in chicken-related samples: in Gambia, *Salmonella* Poona (Dione et al., 2011); in Senegal, *Salmonella* Brancaster (Dione et al., 2009); in Nigeria, *Salmonella* Virchow and *Salmonella* Hiduddify (Raufu et al., 2009; Fashae et al., 2010); and in Cameroon, *Salmonella* Enteritidis (Nzouankeu et al., 2010; Wouafo et al., 2010). These data indicate that there is no specific *Salmonella* serotype typical for chicken. Isolation of *Salmonella* Typhi in our study was of particular significance because this serotype is of human origin and is responsible for typhoid fever, which is still of major concern in developing countries (Kariuki, 2008). *Salmonella* Typhi can be transmitted by the fecal-oral route through contaminated food or water, but its only reservoir is humans. Thus, the detection of *Salmonella* Typhi in a chicken carcass can probably be explained by the poor hygienic practices of the seller.

Salmonella Derby strains isolated in the present study were mostly susceptible to the tested antimicrobials; only five strains were resistant—four to tetracycline and one to streptomycin. None of the strains was resistant to fluoroquinolones. The study conducted in Nigeria found fluoroquinolone-resistant *Salmonella* Derby strains and concluded that they may consti-

tute a public concern because of the presence of the fluoroquinolone-mediating *qnr* genes (Fashae et al., 2010). The location of the *qnr* genes on mobile genetic elements coupled with the indiscriminate use of antimicrobials facilitates selection and the potential spread of resistance genes to other serotypes (Fashae et al., 2010).

The virulence genes that indicate the presence of EPEC were the most prevalent among the chicken carcasses. The EPEC pathogroup usually causes diarrhea in infants, and its prevalence was 16% among diarrheagenic children under 5 years old in Burkina Faso (Bonkoungou et al., 2011). The proportion of EPEC in this study (39%) and in our previous study during the dry season (29%) (Kagambega et al., 2012) was higher than that observed in Cameroon, where 11% of the chickens were contaminated by EPEC (Nzounkeu et al., 2010). In the United States, a very low proportion (1%) of atypical EPEC in chicken breasts was reported (Xia et al., 2010). Only virulence genes carried by atypical EPEC were detected in this study. Atypical EPEC appears to be more closely related to STEC and as such is considered as an emerging pathogen (Trabulsi et al., 2002).

STEC virulence genes were detected in this study in 6% of the chicken carcasses, whereas in our previous study during the dry season, we did not detect any STEC virulence genes among the 30 chicken carcasses studied (Kagambega et al., 2012). The Shiga toxin–encoding genes were also absent from chicken carcasses studied in the United States (Zhao et al., 2001), but in Korea 7% prevalence was detected (Lee et al., 2009). We detected the gene profiles stx_1, eaeA, escV and stx_2, eaeA, escV in chicken carcasses. However, the eaeA gene detected may belong to either STEC or EPEC or both, but this can be confirmed only if the strains are isolated.

EAEC virulence genes were detected in 13% of the chicken carcasses examined, in contrast to the studies conducted in Korea (Lee *et al.*, 2009) and in Burkina Faso during the dry season (Kagambega *et al.*, 2012), where no EAEC was detected in chicken carcasses. In the present study, EAEC virulence genes were detected only in combination with those of EPEC or STEC from the same samples, probably because both pathogroups were present. However, this could also indicate the presence of a so-called hybrid strain that has gained virulence genes from another diarrheagenic *E. coli*, as was the case with the strain that carried genetic determinants of both STEC and EAEC and caused the recent large outbreak in Germany (Mellmann *et al.*, 2011).

The results of our present and previous (Kagambega *et al.*, 2011, 2012) studies suggest that the prevalences of *Salmonella* and diarrheagenic *E. coli* as well as the number of different serotypes and pathogroups present on retail chickens might be higher during the rainy season than during the dry season. Also, the study on the etiology of childhood diarrhea in Burkina Faso suggested that the infections caused by enteropathogenic bacteria were more common during the rainy season (I.J.O. Bonkoungou, personal communication). In any case, the contamination of chickens by *Salmonella* and diarrheagenic *E. coli* was found to be common in Burkina Faso, which raises a public health concern. Therefore, efforts should be made to educate producers, retailers, and consumers on the proper handling and cooking of chicken meat. Follow-up studies should be carried out to monitor the situation in the future.

Acknowledgments

The study was funded by the Academy of Finland grant 122600 for collaboration between the National Institute for Health and Welfare, Finland, and CRSBAN/University of Ouagadougou and by an International Foundation for Science grant to A.K. We thank the personnel of the Bacteriology Unit at the National Institute for Health and Welfare, Finland, for help in serotyping the isolates. The article forms a part of the Ph.D. thesis of A.K.

Disclosure Statement

No competing financial interests exist.

References

Bonkoungou IJO, Lienemann T, Martikainen O, Dembelé R, Sanou I, Traoré AS, Siitonen A, Barro N, and Haukka K. Detection of diarrhoeagenic *Escherichia coli* by 16-plex PCR from young children in urban and rural Burkina Faso. Clin Microbiol Infect 2011; doi: 10.1111/j.1469-0691.2011.03675.x.

Dione MM, Ieven M, Garin B, Marcotty T, and Geerts S. Prevalence and antimicrobial resistance of *Salmonella* isolated from broiler farms, chicken carcasses, and street-vended restaurants in Casamance, Senegal. J Food Prot 2009;72:2423–2427.

Dione MM, Ikumapayi UN, Saha D, Mohammed NI, Geerts S, Ieven M, Adegbola RA, and Antonio M. Clonal differences between non-typhoidal *Salmonella* (NTS) recovered from children and animals living in close contact in the Gambia. PLoS Negl Trop Dis 2011;5:e1148. doi:10.1371/journal.pntd .0001148.

Fashae K, Ogunsola F, Aarestrup FM, and Hendriksen RS. Antimicrobial susceptibility and serovars of *Salmonella* from

chickens and humans in Ibadan, Nigeria. J Infect Dev Ctries 2010;4:484–494.

Kagambega A, Haukka K, Siitonen A, Traoré AS, and Barro N. Prevalence of *Salmonella enterica* and the hygienic indicator *Escherichia coli* in raw meat at markets in Ouagadougou, Burkina Faso. J Food Prot 2011;74:1547–1551.

Kagambega A, Martikainen O, Lienemann T, Siitonen A, Traoré AS, Barro N, and Haukka K. Diarrheagenic *Escherichia coli* detected by 16-plex PCR in raw meat purchased from local markets in Ouagadougou, Burkina Faso. Int J Food Microbiol 2012;153:154–158.

Kariuki S. Typhoid fever in sub-Saharan Africa: challenges of diagnosis and management of infections. J Infect Dev Ctries 2008;2:443–447.

Kauffmann F. Classification and nomenclature of the genus *Salmonella*. Acta Pathol Microbiol Scand [B] Microbiol Immunol 1971;79:421–422.

Lee GY, Jang HI, Hwang IG, and Rhee MS. Prevalence and classification of pathogenic *Escherichia coli* isolated from fresh beef, poultry, and pork in Korea. Int J Food Microbiol 2009; 134:196–200.

Mellmann A, Harmsen D, Cummings CA, Zentz EB, Leopold SR, Rico A, Prior K, Szczepanowski R, Ji Y, Zhang W, McLaughlin SF, Henkhaus JK, Leopold B, Bielaszewska M, Prager R, Brzoska PM, Moore RL, Guenther S, Rothberg JM, and Karch H. Prospective genomic characterization of the German enterohemorrhagic *Escherichia coli* O104:H4 outbreak by rapid next generation sequencing technology. PLoS One 2011;6:e22751.

Nataro JP and Kaper JB. Diarrheagenic *Escherichia coli*. Clin Microbiol Rev 1998;11:142–201.

Nzouankeu A, Ngandjio A, Ejenguele G, Njine T, and Ndayo Wouafo M. Multiple contaminations of chickens with *Campylobacter*, *Escherichia coli* and *Salmonella* in Yaounde (Cameroon). J Infect Dev Ctries 2010;4:583–586.

Raufu I, Hendriksen RS, Ameh JA, and Aarestrup FM. Occurrence and characterization of *Salmonella* Hiduddify from chickens and poultry meat in Nigeria. Foodborne Pathog Dis 2009;6:425–430.

Sackey BA, Mensah P, Collison E, and Sakyi-Dawson E. *Campylobacter*, *Salmonella*, *Shigella* and *Escherichia coli* in live and dressed poultry from metropolitan Accra. Int J Food Microbiol 2000;71:21–28.

Tibaijuka B, Molla B, Hildebrandt G, and Kleer J. Occurrence of *Salmonella* in retail raw chicken products in Ethiopia. Berl Munch Tierarztl Wochenschr 2003;116:55–58.

Todd EC. Epidemiology of foodborne diseases: a worldwide review. World Health Stat 1997;50:30–50.

Trabulsi LR, Keller R, and Tardelli Gomes TA. Typical and atypical enteropathogenic *Escherichia coli*. Emerg Infect Dis 2002;8:508–513.

Varma JK, Molbak K, Barrett TJ, Beebe JL, Jones TF, Rabatsky-Ehr T, Smith KE, Vugia DJ, Chang HG, and Angulo FJ. Antimicrobial-resistant nontyphoidal *Salmonella* is associated with excess bloodstream infections and hospitalizations. J Infect Dis 2005;191:554–561.

White DG, Zhao S, Sulder R, Ayers S, Friedman S, Chen S, McDermott PF, McDermott S, Wagner DD, and Meng J. The isolation of antibiotic-resistant *Salmonella* from retail ground meat. N Engl J Med 2001;345:1147–1154.

Wilson IG. *Salmonella* and *Campylobacter* contamination of raw retail chickens from different producers: a six year survey. Epidemiol Infect 2002;129:635–645.

Wouafo M, Nzouankeu A, Kinfack JA, Fonkoua MC, Ejenguele G, Njine T, and Ngandjio A. Prevalence and antimicrobial resistance of *Salmonella* serotypes in chickens from retail markets in Yaounde (Cameroon). Microb Drug Resist 2010;16:171–176.

Xia X, Meng J, McDermott PF, Ayers S, Blickenstaff K, Tran TT, Abbott J, Zheng J, and Zhao S. Presence and characterization of shiga toxin-producing *Escherichia coli* and other potentially diarrheagenic *E. coli* strains in retail meats. Appl Environ Microbiol 2010;76:1709–1717.

Zhao C, Ge B, De Villena J, Sudler R, Yeh E, Zhao S, White DG, Wagner D, and Meng J. Prevalence of *Campylobacter* spp., *Escherichia coli*, and *Salmonella* serovars in retail chicken, turkey, pork, and beef from the Greater Washington, D.C., area. Appl Environ Microbiol 2001;67:5431–5436.

Address correspondence to:
Assèta Kagambèga, Ph.D.
Laboratoire de Biologie Moléculaire, d'Epidémiologie, et de
Surveillance de Bactéries et Virus Transmis par les Aliments
Centre de Recherche en Sciences Biologiques,
Alimentaires et Nutritionnelles
Département de Biochimie-Microbiologie
UFR-SVT/Université de Ouagadougou
03 B.P. 7021
Ouagadougou 03
Burkina Faso

E-mail: kagamas2007@yahoo.fr

Chapitre 2: *Escherichia coli* **pathogènes dans les fèces de bœufs, poulets et porcs au cours de l'abattage**

I. Introduction

Les *Escherichia coli* font partir des principales bactéries zoonotiques responsables des intoxications alimentaires chez l'homme. Les animaux domestiques, principalement les ruminants sont des réservoirs de ces pathovars (Griffin and Tauxe, 1991). Il existe 5 principaux pathovars de *E. coli* qui sont: *E. coli* entéropathogènes (EPEC), *E. coli* entérotoxinogènes (ETEC), *E. coli* productrices de sigatoxines (STEC), *E. coli* entéroinvasifs (EIEC) et *E. coli* entéroaggrégatifs (EAEC). Vue la proximité des animaux domestiques avec l'homme au Burkina Faso et étant donné qu'ils sont la principale source de viande, cette étude a été menée pour déterminer le niveau de portage des pathogènes de *E. coli* chez le bœuf, le poulet et les porcs.

Pour ce faire, des fèces ont été prélevées au cours de l'abattage de différents animaux provenant de différentes localités du Burkina Faso pour l'isolement de *E. coli*. La détection des 16 gènes de virulence (*uidA, pic, bfpB, invE*, EHEC-*hlyA, elt, ent, escV, eaeA, ipaH, aggR, stx1, stx2, estIa, estIb* et *astA)* des 5 principaux pathovars de *E. coli* a été faite en utilisant la PCR multiplexe.

La prévalence des pathovars a été de 42% chez les bœufs, 49,4% chez les poulets et de 88% chez les porcs. Des infections mixtes, c'est-à-dire plus d'un pathovar dans un même échantillon ont été détectés dans 8% des échantillons de bœufs, dans 7,4% des échantillons de poulets et dans 30% des échantillons de porcs. Les *E. coli* productrices de shiga-toxine (STEC) ont été les plus dominantes au niveau des fèces de bœuf par contre les *E. coli* entéropathogènes (EPEC) ont été les plus dominantes au niveau des poulets. Chez les porcs, la prévalence de STEC et EPEC était la même. *E. coli* entéroinvasive (EIEC) a été détectée seulement au niveau des poulets. Cette bactérie est typique à l'homme et très rarement isolés des animaux. Sa présence chez les poulets peut être expliquée par le faite que ces derniers vivent librement dans les conscessions avec les hommes, se baladent et mangent très souvent les selles.

Les résultats de cette étude montrent que les animaux domestiques au Burkina Faso sont des réservoirs des pathogènes de *E. coli*, ce qui constitue un danger potentiel pour les consommateurs. En plus de la contamination des viandes et carcasses qui peuvent survenir au cours de l'abattage, les fèces de ces animaux sont transportées vers les eaux d'écoulement et de puits. Ce qui favorise un transfert direct de ces pathogènes à l'homme. Ce travail a été publié (**Article 4**).

II.Article 4: Prevalence of diarrheagenic *Escherichia coli* virulence genes in the feces of slaughtered cattle, chickens and pigs in Burkina Faso (*MicrobiologyOpen*, 1(3):276-84. doi: 10.1002/mbo3.30)

MicrobiologyOpen

Prevalence of diarrheagenic *Escherichia coli* virulence genes in the feces of slaughtered cattle, chickens, and pigs in Burkina Faso

Assèta Kagambèga[1,2], Outi Martikainen[1], Anja Siitonen[1], Alfred S. Traoré[2], Nicolas Barro[2] & Kaisa Haukka[1,*]

[1]Bacteriology Unit, Department of Infectious Disease Surveillance and Control, National Institute for Health and Welfare (THL), P.O. Box 30, FI-00271 Helsinki, Finland
[2]Laboratoire de Biologie Moléculaire, d'Epidémiologie et de Surveillance bactéries et virus transmis par les aliments; CRSBAN, Département de Biochimie-Microbiologie, UFR-SVT/Université de Ouagadougou, 03 B.P. 7021, Ouagadougou 03, Burkina Faso

Keywords
Cattle, chickens, diarrheagenic *Escherichia coli*, multiplex PCR, pigs, virulence genes

Correspondence
Assèta Kagambèga, Laboratoire de Biologie Moléculaire, d'Epidémiologie et de Surveillance bactéries et virus transmis par les aliments; CRSBAN, Département de Biochimie-Microbiologie, UFR-SVT/Université de Ouagadougou, 03 B.P. 7021 Ouagadougou 03, Burkina Faso.
Tel: +226 70 24 54 01;
Fax: +226 50 30 72 42;
E-mail: kagamas2007@yahoo.fr

Funding Information
The study was funded by the Academy of Finland grant 122600 to collaboration between the Finnish National Institute for Health and Welfare (THL) and CRSBAN/ University of Ouagadougou.

Received: 27 February 2012; Revised: 28 May 2012; Accepted: 29 May 2012.

doi: 10.1002/mbo3.30

*Current address: Department of Food and Environmental Sciences, P.O. Box 56 FI-00014 University of Helsinki, Finland

Abstract

We investigated the prevalence of the virulence genes specific for five major pathogroups of diarrheagenic *Escherichia coli* (DEC) in primary cultures from feces of animals slaughtered for human consumption in Burkina Faso. For the study, 704 feces samples were collected from cattle ($n = 304$), chickens ($n = 350$), and pigs ($n = 50$) during carcass processing. The presence of the virulence-associated genes in the mixed bacterial cultures was assessed using 16-plex polymerase chain reaction (PCR). Virulence genes indicating presence of DEC were detected in 48% of the cattle, 48% of the chicken, and 68% of the pig feces samples. Virulence genes specific for different DECs were detected in the following percentages of the cattle, chicken, and pig feces samples: Shiga toxin-producing *E. coli* (STEC) in 37%, 6%, and 30%; enteropathogenic *E. coli* (EPEC) in 8%, 37%, and 32%; enterotoxigenic *E. coli* (ETEC) in 4%, 5%, and 18%; and enteroaggregative *E. coli* (EAEC) in 7%, 6%, and 32%. Enteroinvasive *E. coli* (EIEC) virulence genes were detected in 1% of chicken feces samples only. The study was the first of its kind in Burkina Faso and revealed the common occurrence of the diarrheal virulence genes in feces of food animals. This indicates that food animals are reservoirs of DEC that may contaminate meat because of the defective slaughter and storage conditions and pose a health risk to the consumers in Burkina Faso.

Introduction

Animals carry harmless *Escherichia coli* in the intestines as part of the normal gut flora. Sometimes, they are carriers of pathogenic *E. coli* strains that can cause gastrointestinal illness in humans. The importance of these diarrheagenic *E. coli* (DEC) in causing foodborne diseases has been understood in recent years in Africa (Okeke 2009), but very little is known about the reservoirs and routes of the infection on the continent. In general, meat products are considered to be an important source of DEC infections. The meat can be contaminated due to the poor hygiene practices during slaughter. Therefore, adherence to good hygienic practices in slaughter and meat production are essential for prevention of microbial carcass contamination and for ensuring the meat quality and health protection (FAO 2005). Healthy asymptomatic animals may carry zoonotic pathogens and represent a reservoir for DEC, which may enter the food chain via the weak points in hygiene practices of the slaughter process (Hussein

1

2007; Islam et al. 2008; Rhoades et al. 2009). The animals also play an important role in fecal contamination of drinking water sources and agricultural crops enabling direct transfer of zoonotic organisms to humans (Blanco et al. 2003).

At least five main pathogroups of *E. coli* have been associated with human acute intestinal infection. They can be classified based on their virulence genes: Shiga toxin-producing *E. coli* (STEC), enteropathogenic *E. coli* (EPEC), enterotoxigenic *E. coli* (ETEC), enteroaggregative *E. coli* (EAEC), and enteroinvasive *E. coli* (EIEC) (Nataro and Kaper 1998). STEC produces Shiga toxins encoded by *stx1* and/or *stx2* or their variants (Table 1). Besides the *stx* gene (s), STEC strains often carry the *eae* gene, encoding the adherence factor intimin. Also, they often express additional virulence factors, such as enterohemolysin (Schmidt et al. 1995). STEC can cause gastroenteritis that may be complicated by hemorrhagic colitis or hemolytic–uremic syndrome (HUS), which is the main cause of acute renal failure in children (Paton and Paton 1998).

EPEC produces characteristic histopathology known as attaching and effacing (A/E) on intestinal cells (Schmidt 2010) (Table 1). EPEC is further divided into two subtypes, typical (tEPEC) and atypical (aEPEC), depending on the presence or absence of the EPEC adherence factor (EAF) plasmid (Nataro and Kaper 1998; Schmidt 2010). Strains of aEPEC occur frequently also in developed countries, whereas tEPEC is the leading cause of infantile diarrhea in developing countries (Trabulsi et al. 2002). ETEC produces heat-labile (LT) and/or heat-stable (ST) enterotoxins and is an important cause of diarrhea in infants and travelers (Kaper et al. 2004). EIEC is associated with invasive, bloody diarrhea resembling that caused by *Shigella* sp. Invasion is mediated by the genes encoding, for example, Ipa proteins and their transcription regulator *invE* (Nataro and Kaper 1998; Lan et al. 2004). EAEC harbors the mechanism for aggregative-adherence pattern mediated by aggregative-adhesive fimbriae. It is increasingly recognized as a diarrheal pathogen in developing countries (Huang et al. 2004).

Table 1. The *Escherichia coli* pathogroups and their marker genes targeted in the study.

Pathogroup	Gene	Locus	Virulence mechanism	References
STEC	*stx1*	Phage	Shiga toxin 1	Paton and Paton 1998
STEC	*stx2*	Phage	Shiga toxin 2	Paton and Paton 1998
STEC	EHEC-*hly*	Virulence plasmid pO157	Enterohemolysin	Paton and Paton 1998
STEC, EPEC	*eae*	LEE pathogenicity island in the chromosome	Intimin, a protein causing attaching/effacing lesions	Nataro and Kaper 1998
STEC, EPEC	*escV*	LEE pathogenicity island in the chromosome	A conserved area in LEE pathogenicity island	Müller et al. 2007
STEC, EPEC	*ent*	OI-122 pathogenicity island in the chromosome	Enterotoxin or enterohemolysin, a homolog to ShET2 enterotoxin of *Shigella flexneri*l	Müller et al. 2007; Afset et al. 2008
Typical EPEC	*bfpB*	EPEC adherence factor (EAF) plasmid	Subunit of bundle forming pilus (BFP)	Nataro and Kaper 1998; Müller et al. 2007
ETEC	*elt*	Plasmid	Heat-labile enterotoxin LT-I	Nataro and Kaper 1998
ETEC	*estIa*	Plasmid or transposon	Heat-stable enterotoxin ST-Ip	Nataro and Kaper 1998
ETEC	*estIb*	Plasmid or transposon	Heat-stable enterotoxin ST-II	Nataro and Kaper 1998
EIEC	*invE*	Virulence plasmid pINV	Transcription regulator, regulates the *ipa* genes	Hale 1991; Müller et al. 2007
EIEC	*ipaH*	Virulence plasmid pINV and the chromosome	Invasion plasmid antigen	Hale 1991; Hsu et al. 2010
EAEC	*aggR*	Chromosomal island, plasmid pAA	AggR regulon, transcription activator, regulates the genes of fimbrial biogenesis	Cerna et al. 2003; Harrington et al. 2006
EAEC	*pic*	Chromosome	Serine protease	Henderson et al. 1999; Müller et al. 2007
STEC, EPEC, ETEC, EIEC, EAEC	*astA*	Plasmid	EAEC heat-stable enterotoxin (EAST-1)	Nataro and Kaper 1998
STEC, EPEC, ETEC, EIEC, EAEC	*uidA*	Chromosome	β-Glucuronidase	Blanco et al. 1982; Müller et al. 2007

In our previous study on the retail meats at the markets of Ouagadougou, 44% of the beef and 29% of the chicken samples were found to contain DEC virulence genes (Kagambèga et al. 2012). We hypothesized that DEC bacteria might originate especially from fecal contamination of meat during slaughter. In developing countries such as Burkina Faso, the carcass-processing and meat-selling conditions are defective (Kagambèga et al. 2011), which can lead into further proliferation of the pathogens. Recent information is available on the occurrence of DEC in animals in sub-Saharan Africa, such as in Uganda (Majalija et al. 2008), in Ethiopia (Mersha et al. 2010), in Nigeria (Ojo et al. 2010; Akanbi et al. 2011), but little is known about the situation in Burkina Faso. To estimate the risks and the appropriate measures to avoid the risks related to the hygiene of the slaughter process, it is necessary to collect data related to the animals that are shedding potential zoonotic pathogens. This article describes the prevalence of DEC virulence genes in fresh feces collected from slaughtered cattle, chickens, and pigs in Ouagadougou.

Materials and Methods

Sampling

From March to August 2010, we collected altogether 704 fecal samples from cattle ($n = 304$) and pigs ($n = 50$) after slaughter at the central abattoir, and from chickens ($n = 350$) from the local poultry meat sellers in Ouagadougou, Burkina Faso. There were no records available concerning the origin of the animals, but according to the abattoir or the poultry sellers, animals were received from different areas across the country. Immediately after the animals were slaughtered, the rectal material was collected aseptically. The samples were transported to the laboratory and kept at 4°C until the microbiological examination was started within 8 h. Subsequently, of each fecal sample, 25 g was homogenized in 225 mL buffered peptone water (Liofilchem, Teramo, Italy) and enriched at 37°C for 24 h. The enriched samples were cultured on Sorbitol MacConkey (SMAC) agar (Oxoid, Basingstoke, England) at 37°C overnight. Bacterial mass from each plate was collected and stored at −30°C in tubes containing 1 mL of brain heart infusion broth with 15% (v/v) glycerol for further analysis.

16-plex polymerase chain reaction (PCR)

The presence of DEC virulence genes in the feces samples was detected using 16-plex PCR targeting the genes *uidA*, *pic*, *bfpB*, *invE*, EHEC-*hlyA*, *elt*, *ent*, *escV*, *eaeA*, *ipaH*, *aggR*, *stx1*, *stx2*, *estIa*, *estIb*, and *astA* after recultivation the bac-

terial mass on Cystine Lactose Electrolyte Deficient (CLED) agar (Difco, Sparks, USA). The primers, sample preparation, and PCR conditions were as previously described (Antikainen et al. 2009; Kagambèga et al. 2012). The following genes were considered indicative of the *E. coli* pathogroups: for STEC, the presence of *stx1* and/or *stx2* and possibly *eaeA*, *escV*, *ent*, and EHEC-*hly*; for EPEC, the presence of *eaeA* and possibly *escV*, *ent*, and *bfpB*, the absence of *bfpB* indicated atypical EPEC; for ETEC, the presence of *elt* and/or *estIa* and/or *estIb*; for EAEC, the presence of *pic* and/or *aggR*; for EIEC, the presence of *invE* and *ipaH*. The gene *uidA* was used as a general marker for *E. coli*. As *astA* was not specific for a certain pathogroup, it was not utilized in the final analysis.

The following reference strains were used: RH4283 (E 2348/69; Baldini et al. 1983) for EPEC, RH3533 (ATCC 35401) for ETEC, RH4270 (ATCC 43895) for STEC, RH6647 (145-46-215, Statens Serum Institute [SSI], Copenhagen, Denmark) for EIEC, and IH56822 (patient isolate; Keskimäki et al. 2000) for EAEC. The negative controls were *E. coli* RHE6715 (ATCC25922) and sterile distilled water. All the 16-plex PCR positive results were confirmed by single PCRs.

Statistical analysis

The chi-square test or Fisher's exact test of OpenEpi version 2.3.1 were used to determine the statistical significance of the data.

Results

The 16-plex PCR was used to detect the selected virulence genes carried by five pathogroups of *E. coli*. Virulence genes of at least one DEC pathogroup was detected in 348 (49%) of the 704 feces samples, with 149 (21%) being positive for virulence genes of STEC, 172 (24%) of EPEC, 40 (6%) of ETEC, 58 (8%) of EAEC, and 5 (1%) of EIEC (Table 2). Among the different animals, STEC virulence genes were more prevalent in cattle (37%) and pigs (30%) than in chickens (6%). The lower prevalence of STEC virulence genes in chickens than in cattle and pigs was statistically significant ($P < 0.001$). Chickens and pigs had a higher prevalence of EPEC virulence genes (37% and 32%, respectively) than cattle (8%) ($P < 0.001$). ETEC virulence genes were found in 12 (4%), 19 (5%) and EAEC virulence genes in 22 (7%), 20 (6%), and 16 (32%) of cattle, chicken, and pig feces samples, respectively. The higher prevalences of ETEC and EAEC virulence genes in pigs than in cattle or chickens were statistically significant ($P < 0.001$ for both). EIEC virulence genes were found in 5 (1%) of the chicken feces samples. Finding of virulence genes specific for sev-

3

88

Table 2. Diarrheagenic *Escherichia coli* (DEC) pathogroups present in the animal feces, based on the detection of their virulence genes by PCR.

DEC pathogroups	Number of DEC, n (%)			
	Cattle (n = 304)	Chickens (n = 350)	Pigs (n = 50)	Total (n = 704)
Any DEC	145 (48)	169 (48)	34 (68)	348 (49)
STEC	112 (37)	22 (6)	15 (30)	149 (21)
EPEC	25 (8)	131 (37)	16 (32)	172 (24)
ETEC	12 (4)	19 (5)	9 (18)	40 (6)
EAEC	22 (7)	20 (6)	16 (32)	58 (8)
EIEC	0	5 (1)	0	5 (1)

A DEC-positive sample may contain virulence genes of several pathogroups.

eral pathogroups of DEC in samples was common. Virulence genes of more than one pathogroup were detected in 64 (9%) of all the studied samples, in 23 (8%) of the cattle feces samples, in 26 (7%) of the chickens feces samples, and in 15 (30%) of the pig feces samples. However, determination of the copresence of STEC and EPEC virulence genes was not possible using a PCR method, because STEC strains may contain all the virulence genes present in atypical EPEC.

Table 3 shows the occurrence of the 14 genes, which were utilized to analyze the PCR results. Among the 149 samples, which based on the PCR detection were positive for STEC virulence genes, *stx1* without *stx2* was detected in 46 (31%) samples, *stx2* without *stx1* in 36 (24%) samples, and both *stx1* and *stx2* in 67 (45%) samples. Shiga toxin genes together with the genes indicating the presence of LEE pathogenicity island, that is, *eaeA* and/or *escV*, were detected in 53 (36%) of the STEC virulence gene-positive samples. EHEC-*hly* was detected in 34 (23%) of the STEC virulence gene-positive samples and in 3 (2%) of the EPEC virulence gene-positive samples. None of the EPEC virulence gene-positive samples had *bfpB*, so all of them appeared to contain atypical EPEC. Among the 40 ETEC virulence gene positive samples, *estIb* without *estIa* or *elt* was most common with 17 (43%) positive samples, followed by *elt* alone with 10 (25%) positive samples, and *estIa* alone with 6 (15%) positive samples. Five EIEC virulence gene positive samples had *invE*, but no *ipaH*.

Discussion

This study is the first to be undertaken in Burkina Faso to investigate the occurrence of virulence genes specific for DEC in slaughtered animals. The results suggested that animals used for meat production are commonly carriers of the main diarrheagenic pathogroups of *E. coli*. Especially, both STEC and EPEC virulence genes were detected in about a quarter of the samples, whereas ETEC, EAEC, and EIEC virulence genes were less frequently detected.

Of the animals slaughtered for human consumption, 37% of the cattle, 6% of the chicken, and 30% of the pig feces were positive for STEC virulence genes. Also, previously, the ruminants have been indicated as the main natural reservoir for human STEC infections (Caprioli et al. 2005; Cookson et al. 2006; Gyles 2007; Hussein 2007). Investigations on the prevalence of STEC are most commonly based on detection of the O157 serogroup; its detected prevalence in cattle feces varies widely from 0% to over 50% (Rhoades et al. 2009). Based on detection of the *stx* genes, as many as 73% of the healthy cattle in Bangladesh (Islam et al. 2008), 70% in France (Pradel et al. 2000), 69% in Japan (Kobayashi et al. 2001), 25% in Australia (Fagan et al. 1999), and 19% in India (Das et al. 2005) were found to be STEC positive. In Africa, in Nigeria, 10% of cattle were found to be positive for the selected seven STEC serogroups (Ojo et al. 2010).

Analysis of the STEC virulence genes revealed that bovine feces in Burkina Faso mostly harbor both *stx1* and *stx2*, followed by *stx1* alone or *stx2* alone. Similar observations have been made in India (Das et al. 2005) and Germany (Strockbine et al. 1998; Schmidt et al. 1999). In contrast, abundance of *stx2* alone was common in France (Rogerie et al. 2001), Japan (Kobayashi et al. 2001), and Argentina (Blanco et al. 2004). Based on the PCR analysis, it is not possible to know whether the genes detected originate from one or several STEC strains harbored in an animal.

In our study, pigs were found to carry STEC virulence genes nearly as frequently as cattle. It is possible that the freely roaming pigs in Burkina Faso get infected through cattle feces and then serve as an additional possible reservoir for human infections. In Australia, 21% of pig feces tested using PCR were detected to carry *stx* genes (Sidjabat-Tambunan and Bensink 1997). In general, pigs are not considered as major carriers of human pathogenic STEC, although some STEC O157 strains from pigs have been isolated (Heuvelink et al. 1999; Johnsen et al. 2001; Bonardi et al. 2003; Caprioli et al. 2005). However, F18 fimbriae-positive *E. coli* producing the Shiga toxin variant

4

Table 3. Number of virulence genes detected by 16-plex PCR in 304 cattle, 350 chicken, and 50 pig feces samples and in the six control strains.

Pathogroups		stx1	stx2	hly	eaeA	escV	ent	bfpB	elt	estIa	estIb	aggR	pic	ipaH	invE
	Control strains														
STEC	RH4270	+	+	+	+	+	+	–	–	–	–	–	–	–	–
EPEC	RH4283	–	–	–	+	+	+	–	–	–	–	–	–	–	–
ETEC	RH3533	–	–	–	–	–	–	–	+	–	+	–	–	–	–
EAEC	IH56822	–	–	–	–	–	–	–	–	–	–	+	+	–	–
EIEC	RH6647	–	–	–	–	–	–	–	–	–	–	–	–	+	+
Negative control	RH6715	–	–	–	–	–	–	–	–	–	–	–	–	–	–
	Cattle*														
STEC	93	73	68	18	10	14	5	–	–	–	–	–	–	–	–
EPEC	22	–	–	1	11	21	11	–	–	–	–	–	–	–	–
EAEC	7	–	–	–	–	–	–	–	–	–	–	4	4	–	–
STEC + ETEC	7	5	6	2	2	1	–	–	2	1	5	–	–	–	–
STEC + EAEC	9	6	7	2	2	3	1	–	–	–	–	6	3	–	–
EPEC + ETEC	1	–	–	–	1	1	–	–	–	1	–	–	–	–	–
EPEC + EAEC	2	–	–	–	–	2	1	–	–	–	–	2	–	–	–
ETEC + EAEC	1	–	–	–	–	–	–	–	1	–	–	–	1	–	–
STEC + EAEC + ETEC	3	3	3	–	2	–	1	–	–	–	3	2	2	–	–
	Chickens*														
STEC	15	10	7	3	3	10	2	–	–	–	–	–	–	–	–
EPEC	113	–	–	1	66	108	31	–	–	–	–	–	–	–	–
ETEC	6	–	–	–	–	–	–	–	4	1	2	–	–	–	–
EAEC	4	–	–	–	–	–	–	–	–	–	–	2	2	–	–
EIEC	5	–	–	–	–	–	–	–	–	–	–	–	–	5	–
STEC + ETEC	4	4	–	1	–	3	–	–	1	3	–	–	–	–	–
STEC + EAEC	3	1	2	–	1	1	1	–	–	–	–	3	–	–	–
EPEC + ETEC	6	–	–	–	2	5	2	–	2	4	–	–	–	–	–
EPEC + EAEC	10	–	–	–	4	10	3	–	–	–	–	9	1	–	–
ETEC + EAEC	1	–	–	–	–	–	–	–	1	–	–	–	1	–	–
EPEC + EAEC + ETEC	2	–	–	–	2	2	1	–	–	1	1	1	1	–	–
	Pigs*														
STEC	9	6	7	5	–	5	1	–	–	–	–	–	–	–	–
EPEC	7	–	–	–	6	7	2	–	–	–	–	–	–	–	–
EAEC	3	–	–	–	–	–	–	–	–	–	–	2	1	–	–
STEC + ETEC	2	2	1	2	1	2	–	–	1	–	1	–	–	–	–
STEC + EAEC	2	2	1	1	1	1	–	–	–	–	–	2	–	–	–
EPEC + EAEC	4	–	–	1	3	4	1	–	–	–	–	4	–	–	–
STEC + EAEC + ETEC	2	1	1	–	–	2	–	–	2	–	1	2	–	–	–
EPEC + EAEC + ETEC	5	–	–	–	5	5	–	–	4	2	2	5	–	–	–

*Number of the positive cattle/chicken/pig samples for each pathogroup or pathogroup combination.

Stx2e are significant pathogens of pigs (da Silva et al. 2001).

In chickens, we detected 6% prevalence of STEC virulence genes. This is comparable with 9% of STEC found from chicken feces in Tanzania (Raji et al. 2006). However, in several other studies no STEC was detected in poultry (Heuvelink et al. 1999; Kobayashi et al. 2002) or the prevalence was low compared with that in cattle (Schouten et al. 2005; Dipineto et al. 2006).

We found EPEC virulence genes to be more prevalent in chicken (37%) and pig (32%) feces than in cattle feces

(8%). These numbers are probably an underestimate, as several samples might actually have contained EPEC in addition to STEC. It was not possible to detect their coinfection by a direct PCR technique. However, the 37% prevalence of EPEC virulence genes in chicken feces was in line with the previous reports where retail chickens were found to be commonly contaminated by EPEC (Lee et al. 2009); we found 29% prevalence in Burkina Faso (Kagambèga et al. 2012). In Argentina, up to 26% of cloacal swabs of chicken and 58% of the slaughtered chicken carcasses were positive for EPEC (Alonso et al. 2011).

The 8% prevalence of EPEC virulence genes in cattle feces in this study was lower than 31% found in cattle feces in New Zealand (Cookson et al. 2006). The prevalence of EPEC virulence genes in pigs (32%) was higher than 18% found in pigs in Germany (Krause et al. 2005). Only virulence genes of atypical EPEC were detected in this study. Typical EPEC is, indeed, rarely isolated from animals, whereas atypical EPEC strains are isolated from both animals and humans (Nataro and Kaper 1998; Aktan et al. 2004; Krause et al. 2005). In our previous study among children in Burkina Faso, atypical EPEC were detected in the feces of diarrheagenic children more often than typical EPEC (Bonkoungou et al., in press).

ETEC and EAEC virulence genes were found at lower rates than STEC and EPEC virulence genes from the feces of the studied animals with an exception of 32% prevalence of EAEC virulence genes in pigs. In a previous study in Brazil (Uber et al. 2006), the EAEC isolates from calves, piglets, and horses were found to differ genetically from the human isolates. The strains of animal origin lacked *aggR*, but harbored other marker genes, such as *pic*. Among the feces samples in this study, *aggR* was more common than *pic*. Furthermore, in our previous study, we found *aggR* from raw meat samples, but never *pic* (Kagambèga et al. 2012).

ETEC is a major cause of severe diarrheal disease in suckling and weanling animals. For the pig industry, ETEC diarrhea is causing considerable losses also in Africa. In Zimbabwe, 32% of the studied piglets tested positive for STa, STb, LT, or Stx-2e genes (Madoroba et al. 2009). Animal-derived strains are known to produce enterotoxins similar to those of human strains, but colonization factors necessary for colonization of the host small bowel are species specific and different in human and animal strains (Qadri et al. 2005). In our study, *estIb* was found to be most prevalent among the 40 ETEC virulence gene-positive samples followed by samples positive for *elt* and *estIa*. In our previous study on raw meat samples, we detected different proportions of these genes, *elt* being the most common and *estIb* the least common (Kagambèga et al. 2012).

EIEC virulence genes were detected only in chicken feces in this study. However, detection of the EIEC-specific genes *ipaH* and *invE* can also indicate the presence of *Shigella* sp. in the sample. For both EIEC and *Shigella*, the reservoir is considered to be the gut of infected humans (Meng et al. 2007). Finding of this pathogroup in chicken feces might be explained by the typical close contact of the chickens and humans in developing countries, where chicken roam freely on the yards and the surroundings.

The role of pigs as a reservoir for diarrheal DEC has not been commonly recognized. However, in this study, we found rather high occurrence of STEC, EPEC, and EAEC virulence genes in pig feces. Furthermore, virulence genes specific for more than one DEC pathogroup were mostly found among the pigs. Therefore, pigs may be a noteworthy reservoir for several DEC pathogroups in Burkina Faso. However, comparison of isolates from both animal and human sources would be needed to evaluate the zoonotic risk for humans. Yet, in future studies, more attention should be given to pigs as a potential source of zoonotic DEC infections, because in developing countries they often live in close contact with humans. The results from the previous studies, where the similarity of the *E. coli* isolates from humans and animals has been investigated are somewhat ambiguous (Cookson et al. 2010; Clermont et al. 2011). In the studies conducted in Africa, Kariuki et al. (1999) concluded that although several different pulsed-field gel electrophoresis genotypes of *E. coli* were isolated from children and chickens from the same farms in Kenya, the *E. coli* strains from these two sources were distinct. Rwego et al. (2008) ended up in quite the opposite conclusion in Uganda, suggesting that both rates of human–livestock interactions and patterns of human hygiene affect human–livestock bacterial transmission in the rural setting they studied.

The high prevalence of DEC virulence genes detected in this study suggests widespread occurrence of DEC in the feces of slaughtered cattle, chickens, and pigs in Burkina Faso. This finding together with the defective meat retail conditions creates a potential infection route for foodborne pathogens. Good hygienic practices at slaughterhouses and processing plants as well as at home are necessary to minimize the risk of human DEC infections. The multiplex PCR approach is well suited for rapid and sensitive detection of the presence of the DEC virulence genes.

Acknowledgments

The study was funded by the Academy of Finland grant 122600 to collaboration between the Finnish National Institute for Health and Welfare (THL) and CRSBAN/ University of Ouagadougou.

Conflict of Interest

None declared.

References

Afset, J. E., E. Anderssen, G. Bruant, J. Harel, L. Wieler, and K. Bergh. 2008. Phylogenetic backgrounds and virulence profiles of atypical enteropathogenic *Escherichia coli* strains from a case–control study using multilocus sequence typing and DNA microarray analysis. J. Clin. Microbiol. 46:2280–2290.

Akanbi, B. O., I. P. Mbah, and P. C. Kerry. 2011. Prevalence of *Escherichia coli* O157:H7 on hides and faeces of ruminants at slaughter in two major abattoirs in Nigeria. Lett. Appl. Microbiol. 53:336–240.

Aktan, I., K. A. Sprigings, R. M. La Ragione, L. M. Faulkner, G. A. Paiba, and M. J. Woodward. 2004. Characterization of attaching-effacing *Escherichia coli* isolated from animals at slaughter in England and Wales. Vet. Microbiol. 102:43–53.

Alonso, M. Z., N. L. Padola, A. E. Parma, and P. M. A. Lucchesi. 2011. Enteropathogenic *Escherichia coli* contamination at different stages of the chicken slaughtering process. Poult. Sci. 90:2638–2641.

Antikainen, J., E. Tarkka, K. Haukka, A. Siitonen, M. Vaara, and J. Kirveskari. 2009. New 16-plex PCR method for rapid detection of diarrheagenic *Escherichia coli* directly from stool samples. Eur. J. Clin. Microbiol. Infect. Dis. 28:899–908.

Baldini, M. M., J. B. Kaper, M. M. Levine, D. C. Candy, and H. W. Moon. 1983. Plasmid-mediated adhesion in enteropathogenic *Escherichia coli*. J. Pediatr. Gastroenterol. Nutr. 2:534–538.

Blanco, C., P. Ritzenthaler, and M. Mata-Gilsinger. 1982. Cloning and endonuclease restriction analysis of *uidA* and *uidR* genes in *Escherichia coli* K-12: determination of transcription direction for the *uidA* gene. J. Bacteriol. 149:587–594.

Blanco, J., M. Blanco, J. E. Blanco, A. Mora, E. Gonzales, M. I. Bernardez, et al. 2003. Verotoxin-producing *Escherichia coli* in Spain: prevalence, serotypes and virulence genes of O157:H7 and non-O157 VTEC in ruminants, raw beef products and humans. Exp. Biol. Med. 228:345–351.

Blanco, M., N. L. Padola, A. Krüger, M. E. Sanz, J. E. Blanco, E. A. González, et al. 2004. Virulence genes and intimin types of Shiga-toxin-producing *Escherichia coli* isolated from cattle and beef products in Argentina. Int. Microbiol. 7:269–276.

Bonardi, S., F. Brindani, G. Pizzin, L. Lucidi, M. D'Incau, E. Liebana, et al. 2003. Detection of *Salmonella* spp., *Yersinia enterocolitica* and verocytotoxin-producing *Escherichia coli* O157 in pigs at slaughter in Italy. Int. J. Food Microbiol. 15:101–110.

Bonkoungou, I. J., T. Lienemann, O. Martikainen, R. Dembelé, I. Sanou, A. S. Traoré, et al. In press. Diarrhoeagenic *Escherichia coli* detected by 16-plex PCR in children with and without diarrhoea in Burkina Faso. Clin. Microbiol. Infect. doi: 10.1111/j.1469-0691.2011.03675.x

Caprioli, A., S. Morabito, H. Brugère, and E. Oswald. 2005. Enterohaemorrhagic *Escherichia coli*: emerging issues on virulence and modes of transmission. Vet. Res. 36:289–311.

Cerna, J. F., J. P. Nataro, and T. Estrada-Garcia. 2003. Multiplex PCR for detection of three plasmid-borne genes of enteroaggregative *Escherichia coli* strains. J. Clin. Microbiol. 41:2138–2140.

Clermont, O., M. Olier, C. Hoede, L. Diancourt, S. Brisse, M. Keroudean, et al. 2011. Animal and human pathogenic *Escherichia coli* strains share common genetic backgrounds. Infect. Genet. Evol. 11:654–662.

Cookson, A. L., S. C. S. Taylor, J. Bennett, F. Thomson-Carter, and G. T. Attwood. 2006. Serotypes and analysis of distribution of Shiga toxin-producing *Escherichia coli* from cattle and sheep in the lower North Island, New Zealand. N Z Vet. J. 54:78–84.

Cookson, A. L., M. Cao, J. Bennett, C. Nicol, F. Thomson-Carter, and G. T. Attwood. 2010. Relationship between virulence gene profiles of atypical enteropathogenic *Escherichia coli* and Shiga toxin-producing *E. coli* isolates from cattle and sheep in New Zealand. Appl. Environ. Microbiol. 76:3744–3747.

Das, S. C., A. Khan, P. Panja, S. Datta, A. Sikdar, S. Yamasaki, et al. 2005. Dairy farm investigation on Shiga toxin-producing *Escherichia coli* (STEC) in Kolkata, India with emphasis on molecular characterization. Epidemiol. Infect. 133:617–626.

Dipineto, L., A. Santaniello, M. Fontanella, K. Lagos, A. Fioretti, and L. F. Menna. 2006. Presence of Shiga toxin-producing *Escherichia coli* O157:H7 in living layer hens. Lett. Appl. Microbiol. 43:293–295.

Fagan, P. K., M. A. Hornitzky, K. A. Bettelheim, and S. P. Djordjevic. 1999. Detection of Shiga-like toxin (*stx1* and *stx2*), intimin (*eaeA*), and enterohemorrhagic *Escherichia coli* (EHEC) hemolysin (EHEC *hlyA*) genes in animal feces by multiplex PCR. Appl. Environ. Microbiol. 65:868–872.

FAO, Food and Agriculture Organization. 2005. Code of hygienic practice for meat. CAC/RCP 58–200. Codex Alimentarius, FAO, Rome.

Gyles, C. L. 2007. Shiga toxin-producing *Escherichia coli*: an overview. J. Anim. Sci. 85:45–62.

Hale, T. L. 1991. Genetic basis of virulence in *Shigella* species. Microbiol. Rev. 55:206–224.

Harrington, S. M., E. G. Dudley, and J. P. Nataro. 2006. Pathogenesis of enteroaggregative *Escherichia coli* infection. FEMS Microbiol. Lett. 254:12–18.

Henderson, I. R., J. Czeczulin, C. Eslava, F. Noriega, and J. P. Nataro. 1999. Characterization of Pic, a secreted protease of *Shigella flexneri* and enteroaggregative *Escherichia coli*. Infect. Immun. 67:5587–5596.

Heuvelink, A. E., J. T. Zwartkruis-Nahuis, F. L. van den Biggelaar, W. J. van Leeuwen, and E. de Boer. 1999. Isolation and characterization of verocytotoxin-producing *Escherichia coli* O157 from slaughter pigs and poultry. Int. J. Food Microbiol. 52:67–75.

Hsu, B. M., S. F. Wu, S. W. Huang, Y. J. Tseng, D. D. Ji, J. S. Chen, et al. 2010. Differentiation and identification of *Shigella* spp. and enteroinvasive *Escherichia coli* in environmental waters by a molecular method and biochemical test. Water Res. 44:949–955.

Huang, D. B., P. C. Okhuysen, Z. Jiang, and H. L. Dupont. 2004. Enteroaggregative *Escherichia coli*: an emerging enteric pathogen. Am. J. Gastroenterol. 99:383–389.

Hussein, H. S. 2007. Prevalence and pathogenicity of shiga toxin-producing *Escherichia coli* in beef cattle and their products. J. Anim. Sci. 85:63–72.

Islam, M. A., A. S. Mondol, E. de Boer, R. R. Beumer, M. H. Zwietering, K. A. Talukder, et al. 2008. Prevalence and genetic characterization of Shiga toxin-producing *Escherichia coli* isolates from slaughtered animals in Bangladesh. Appl. Environ. Microbiol. 74:5414–5421.

Johnsen, G., Y. Wasteson, E. Heir, O. I. Berget, and H. Herikstad. 2001. *Escherichia coli* O157:H7 in faeces from cattle, sheep and pigs in the southwest part of Norway during 1998 and 1999. Int. J. Food Microbiol. 65:193–200.

Kagambèga, A., K. Haukka, A. Siitonen, A. S. Traoré, and N. Barro. 2011. Prevalence of *Salmonella enterica* and the hygienic indicator *Escherichia coli* in raw meat at markets in Ouagadougou, Burkina Faso. J. Food Prot. 74:1547–1551.

Kagambèga, A., O. Martikainen, T. Lienemann, A. Siitonen, A. S. Traoré, N. Barro, et al. 2012. Diarrheagenic *Escherichia coli* detected by 16-plex PCR in raw meat and beef intestines sold at local markets in Ouagadougou, Burkina Faso. Int. J. Food Microbiol. 153:154–158.

Kaper, J. B., J. P. Nataro, and H. L. Mobley. 2004. Pathogenic *Escherichia coli*. Nat. Rev. Microbiol. 2:123–1240.

Kariuki, S., C. Gilks, J. Kimari, A. Obanda, J. Muyodi, P. Waiyaki, et al. 1999. Genotype analysis of *Escherichia coli* strains isolated from children and chickens living in close contact. Appl. Environ. Microbiol. 65:472–476.

Keskimäki, M., L. Mattila, H. Peltola, and A. Siitonen. 2000. Prevalence of diarrheagenic *Escherichia coli* in Finns with or without diarrhea during a round-the-world trip. J. Clin. Microbiol. 38:4425–4429.

Kobayashi, H., J. Shimada, M. Nakazawa, T. Morozumi, T. Pohjanvirta, S. Pelkonen, et al. 2001. Prevalence and characteristics of Shiga toxin-producing *Escherichia coli* from healthy cattle in Japan. Appl. Environ. Microbiol. 67:484–489.

Kobayashi, H., T. Pohjanvirta, and S. Pelkonen. 2002. Prevalence and characteristics of intimin and Shiga toxin-producing *Escherichia coli* from gulls, pigeons and broilers in Finland. J. Vet. Med. Sci. 64:1071–1073.

Krause, G., S. Zimmermann, and L. Beutin. 2005. Investigation of domestic animals and pets as a reservoir for intimin- (eae) gene positive *Escherichia coli* types. Vet. Microbiol. 106:87–95.

Lan, R., M. A. Alles, K. Donohoe, M. B. Martinez, and P. R. Reeves. 2004. Molecular evolutionary relationships of enteroinvasive *Escherichia coli* and *Shigella* spp. Infect. Immun. 72:5080–5088.

Lee, G. Y., H. I. Jang, I. G. Hwang, and M. S. Rhee. 2009. Prevalence and classification of pathogenic *Escherichia coli* isolated from fresh beef, poultry, and pork in Korea. Int. J. Food Microbiol. 134:196–200.

Madoroba, E., E. Van Driessche, H. De Greve, J. Mast, I. Ncube, J. Read, et al. 2009. Prevalence of enterotoxigenic *Escherichia coli* virulence genes from scouring piglets in Zimbabwe. Trop. Anim. Health Prod. 41:1539–1547.

Majalija, S., H. Segal, F. Ejobi, and B. G. Elisha. 2008. Shiga toxin gene-containing *Escherichia coli* from cattle and diarrheic children in the pastoral systems of southwestern Uganda. J. Clin. Microbiol. 46:352–354.

Meng, J., M. P. Doyle, T. Zhao, and S. Zhao. 2007. Enterohemorrhagic *Escherichia coli*. Pp. 249–269 in M. P. Doyle and L. R. Beuchat, eds. Food microbiology: fundamentals and frontiers. 3rd ed. American Society for Microbiology, Washington, DC.

Mersha, G., D. Asrat, B. M. Zewde, and M. Kyule. 2010. Occurrence of *Escherichia coli* O157:H7 in faeces, skin and carcasses from sheep and goats in Ethiopia. Lett. Appl. Microbiol. 50:71–76.

Müller, D., L. Greune, G. Heusipp, H. Karch, A. Fruth, H. Tschäpe, et al. 2007. Identification of unconventional intestinal pathogenic *Escherichia coli* isolates expressing intermediate virulence factor profiles by using a novel single-step multiplex PCR. Appl. Environ. Microbiol. 73:3380–3390.

Nataro, J. P., and J. B. Kaper. 1998. Diarrheagenic *Escherichia coli*. Clin. Microbiol. Rev. 11:142–201.

Ojo, O. E., A. T. P. Ajuwape, E. B. Otesile, A. A. Owoade, M. A. Oyekunle, and A. I. Adetosoye. 2010. Potentially zoonotic shiga toxin-producing *Escherichia coli* serogroups in the faeces and meat of food-producing animals in Ibadan, Nigeria. Int. J. Food Microbiol. 142:214–221.

Okeke, I. N. 2009. Diarrheagenic *Escherichia coli* in sub-Saharan Africa: status, uncertainties and necessities. J. Infect. Dev. Ctries. 3:817–842.

Paton, J. C., and A. W. Paton. 1998. Pathogenesis and diagnosis of Shiga toxin-producing *Escherichia coli* infections. Clin. Microbiol. Rev. 11:450–479.

Pradel, N., V. Livrelli, C. D. De Champs, J. B. Palcoux, A. Reynaud, F. Scheutz, et al. 2000. Prevalence and characterization of Shiga toxin-producing *Escherichia coli* isolated from cattle, food, and children during a one-year prospective study in France. J. Clin. Microbiol. 38:1023–1031.

Qadri, F., A. M. Svennerholm, A. S. G. Faruque, and R. B. Sack. 2005. Enterotoxigenic *Escherichia coli* in developing countries: epidemiology, microbiology, clinical features, treatment, and prevention. Clin. Microbiol. Rev. 18:465–483.

Raji, M. A., U. M. Minga, and R. S. Machang'u. 2006. Prevalence and characterization of verotoxigenic *Escherichia coli* O157 isolated from local chickens in Morogoro, Tanzania. J. Anim. Vet. Adv. 5:952–958.

Rhoades, J. R., G. Duffy, and K. Koutsoumanis. 2009. Prevalence and concentration of verocytotoxigenic *Escherichia coli*, *Salmonella enterica* and *Listeria monocytogenes* in the beef production chain: a review. Food Microbiol. 26:357–376.

Rogerie, F., A. Marecat, S. Gambade, F. Dupond, P. Beaubois, and M. Lange. 2001. Characterization of Shiga toxin producing *E. coli* and O157 serotype *E. coli* isolated in France from healthy domestic cattle. Food Microbiol. 63:217–223.

Rwego, I. B., T. R. Gillespie, G. Isabirye-Basuta, and T. L. Goldberg. 2008. High rates of *Escherichia coli* transmission between livestock and humans in rural Uganda. J. Clin. Microbiol. 46:3187–3191.

Schmidt, M. A. 2010. LEE ways: tales of EPEC, ATEC and EHEC. Cell. Microbiol. 12:1544–1552.

Schmidt, H., L. Beutin, and H. Karch. 1995. Molecular analysis of the plasmid-encoded hemolysin of *Escherichia coli* O157: H7 strain EDL 933. Infect. Immun. 63:1055–1061.

Schmidt, H., C. Geitz, I. T. Phillips, F. Mathias, and H. Karch. 1999. Non-O157 pathogenic Shiga toxin producing *Escherichia coli*: Phenotypic and genetic profiling of virulence traits and evidence for clonality. J. Infect. Dis. 179:115–123.

Schouten, J. M., A. W. van de Giessen, K. Frankena, M. C. De Jong, and E. A. Graat. 2005. *Escherichia coli* O157 prevalence in Dutch poultry, pig finishing and veal herds and risk factors in Dutch veal herds. Prev. Vet. Med. 70:1–15.

Sidjabat-Tambunan, H., and J. C. Bensink. 1997. Verotoxin-producing *Escherichia coli* from the faeces of sheep, calves and pigs. Aust. Vet. J. 75:292–293.

da Silva, A. S., G. F. Valadares, M. P. Penatti, B. G. Brito, and D. da Silva Leite. 2001. *Escherichia coli* strains from edema disease: O serogroups, and genes for Shiga toxin, enterotoxins, and F18 fimbriae. Vet. Microbiol. 6:227–233.

Strockbine, N. A., J. G. Wells, C. A. Bopp, and T. J. Barrett. 1998. Overview of detection and subtyping methods. Pp. 331–356 in J. B. Kaper and A. D. O'Brien, eds. *Escherichia coli* O157:H7 and other Shiga toxin producing *E. coli* strains. American Society for Microbiology, Washington, DC.

Trabulsi, L. R., R. Keller, and T. A. Tardelli Gomes. 2002. Typical and atypical enteropathogenic *Escherichia coli*. Emerg. Infect. Dis. 8:508–513.

Uber, A. P., L. R. Trabulsi, K. Irino, L. Beutin, A. C. R. Ghilardi, T. A. T. Gomes, et al. 2006. Enteroaggregative *Escherichia coli* from humans and animals differ in major phenotypical traits and virulence genes. FEMS Microbiol. Lett. 256:251–257.

9

Chapitre 3: Caracterisation moléculaire de *Salmonella* dans les fèces de bœufs, poulets et porcs au cours de l'abattage et chez le hérisson

I. Introduction

Les animaux domestiques et de compagnie sont considérés comme une source principale des salmonelloses humaines (Blanco *et al.*, 2004; Woodward *et al.*, 1997). Au cours de nos investigations sur les viandes, nous avons identifié sur la viande et boyaux de bœuf un sérotype de *Salmonella* qui est *S*. Tilene dont l'origine avait été déjà attribuée aux hérisssons d'Afrique par des études antérieures (Lipsky et Tanino, 1995). Ce qui nous a amenés à supposer des contaminations croisées entre herbe et fèces de hérisson au cours du patûrage et/ou de l'abreuvage des bœufs. Dans cette présente étude, nous avons inclus dans l'échantillonage des fèces d'animaux d'abattages, le hérisson pour déterminer leur part dans les Salmonelloses humaines.

Au Burkina Faso, le hérisson est un animal sauvage chassé pour sa viande dans beaucoup de villages et il est aussi exporté dans les pays développés comme un animal de compagnie. Une surveillance des sources et de la distribution des sérotypes de *Salmonella* au niveau des différents animaux qui sont source de viande et/ou de lait et de certains animaux de compagnie sont nécessaires pour réduire l'expansion de ces pathogènes et les infections liées aux aliments d'origine animale. Le présent travail a pour objectif premier, de déterminer la prévalence des sérotypes de *salmonella* chez les animaux domestiques et de vérifier aussi la part zoonotique des hérissons dans les salmonelloses humaines. Le second objectif est de déterminer la sensibilité aux antibiotiques des sérotypes isolés et de comparer génétiquement les sérotypes identiques d'un même et/ou différents échantillons avec ceux isolés des selles de patients malades du Burkina Faso par l'électrophorèse en champ pulsé (PFGE). Les fèces des différents animaux ont été collectées lors de l'abattage pour les différentes analyses.

Au total, 383 *Salmonella* constituées de 75 différents sérotypes ont été identifiées avec 159 (52%) chez les bœufs, 192 (55%) chez les poulets, 8 (16%) chez les porcs et 24 (96%) chez les hérissons. La résistance à certains antibiotiques a été de 14% au niveau des isolats. L'électrophorèse en champ pulsé à révélé des degrés de similitude allant de 75-99% entre les souches. Par exemple les souches de *S.* Typhimurium isolées des fèces du poulet et des selles des patients malades ont eu le même profil de résistance aux antibiotiques et presentaient un degré de similitude de 90-95%. Cette étude révèle la nécessité d'un programme de surveillance des *Salmonella* au niveau national vu que les mêmes sérotypes circulent entre différents animaux et l'homme. Les résultats de cette étude ont fait l'objet d'un projet d'article à soumettre pour publication (**Article 5**).

II. Article 5: Antimicrobial susceptibility, PFGE and serovars of *Salmonella* from cattle, poultry, pigs, hedgehogs and humans in Burkina Faso. (soumis).

Prevalence and characterization of *Salmonella enterica* from the feces of cattle, poultry, swine and hedgehogs in Burkina Faso and their comparison to human *Salmonella* isolates

Asséta Kagambèga[1,2*], Taru Lienemann[1], Laura Aulu[1], Alfred S. Traoré[2], Nicolas Barro[2], Anja Siitonen[1] & Kaisa Haukka [1,3]

[1] Bacteriology Unit, Department of Infectious Disease Surveillance and Control, National Institute for Health and Welfare (THL), Helsinki, Finland

[2] Laboratoire de Biologie Moléculaire et d'Epidémiologie et de Surveillance Bactéries et Virus transmis par les Aliments : CRSBAN, Département de Biochimie-Microbiologie, UFR-SVT/Université de Ouagadougou, Ouagadougou, Burkina Faso

[3] Department of Food and Environmental Sciences, Division of Microbiology, University of Helsinki, Finland

* Author for correspondence

E-mail addresses:

AK: kagamas2007@yahoo.fr

TL: taru.lienemann@thl.fi

LA: laura.aulu@gmail.com

AST: astraore@yahoo.fr

NB: barronicolas@yahoo.fr

AS: anja.siitonen@thl.fi

KH: kaisa.haukka@helsinki.fi

Abstract

Background: Production and wild animals are major sources of human salmonellosis and animals raised for food also play an important role in transmission of antimicrobial resistant *Salmonella* strains to humans. Furthermore, in sub-Saharan Africa non-typhoidal *Salmonella* serotypes are common bloodstream isolates in febrile patients. Yet, little is known about the environmental reservoirs and predominant modes of transmission of these pathogens. The purpose of this study was to discover potential sources and distribution vehicles of *Salmonella* by isolating strains from apparently healthy slaughtered food animals and wild hedgehogs and by determining the genetic relatedness between the strains and human isolates. For this purpose, 729 feces samples from apparently healthy slaughtered cattle (n=304), poultry (n=350), swine (n=50) and hedgehogs (n=25) were examined for the presence of *Salmonella enterica* in Burkina Faso. The isolates were characterized by serotyping, antimicrobial-susceptibility testing, phage typing, and pulsed-field gel electrophoresis (PFGE) with *Xba*I and *Bln*I restriction enzymes.

Results: Of the 729 feces samples, 383 (53%) contained *Salmonella*, representing a total of 81 different serotypes. *Salmonella* was present in 52% of the cattle, 55% of the poultry, 16% of the swine and 96% of the hedgehog feces samples. Antimicrobial resistance was detected in 14% of the isolates. *S.* Typhimurium isolates from poultry and humans were multiresistant to the same antimicrobials (ampicillin, chloramphenicol, streptomycin, sulfonamides and trimethoprim), had the same phage type DT 56 and were closely related in PFGE. *S.* Muenster isolates from hedgehogs had similar PFGE patterns as the domestic animals.

Conclusions: Based on our results it seems that production and wild animals can share the same *Salmonella* serotypes and potentially transmit some of them to humans. As the humans and animals often live in close vicinity in Africa and the hygiene control of the meat retail chain is defective, high *Salmonella*

2

are   another

carriage rates of the animals can pose a major public health risk in Burkina Faso. This underlines the necessity for a joint and coordinated surveillance and monitoring programs for salmonellosis in Africa.

Keywords: *Salmonella*, serotypes, antimicrobial resistance, genetic relatedness, PFGE

3

Background

Salmonella is one of the major zoonotic foodborne pathogens worldwide. It can cause a variety of clinical manifestations from mild gastroenteritis to bacteremia and typhoid fever. The global burden of nontyphoidal *Salmonella* gastroenteritis has been estimated to be 93.8 million cases of gastroenteritis each year, with 155 000 deaths [1]. In Africa, non-typhoidal *Salmonella* has consistently been reported as a leading cause of bacteremia among immuno-compromised people, infants and newborns [2,3]. However, the sources and transmission routes of *Salmonella* in developing countries are poorly understood due to the lack of coordinated national epidemiological surveillance systems [4,5]. In general, the primary sources of salmonellosis are considered to be food-producing animals such as cattle, poultry and swine [6]. The pathogens are mainly disseminated by trade in animals and uncooked animal food products [7]. The process of removing the gastrointestinal tract during slaughtering of food animals is regarded as one of the most important sources of carcass and organ contamination with *Salmonella* at abattoirs [8]. Also asymptomatic pet animals are a potential source of infection, especially species with high fecal carriage rates of *Salmonella* [9]. African pygmy hedgehogs kept as pets have previously been associated with cases of human salmonellosis [10]. The development and the accumulation of resistance to antimicrobials in foodborne pathogens are a major problem for public health. Multi-resistant *Salmonella* may acquire their resistance genes from microbiota of production animals before being transmitted to humans through food chain [11,12].

Due to the lacking surveillance programs in Burkina Faso, as in the most of Africa, information on the prevalence of *Salmonella* and other enteropathogens in food stuffs is limited. However, our previous study on the prevalence of enteric bacteria on retail meats sold at the markets in Ouagadougou, Burkina Faso, revealed that 37% of the chicken, 13% of the beef intestines, and 7% of the mutton samples were contaminated by *Salmonella* [13]. The most common serotypes detected were *S.* Derby and *S.* Tilene. In a

4

following broader study on chicken carcasses in Burkina Faso, up to 57% of the carcasses were found to be contaminated by *Salmonella*. *S*. Derby again being the most common serotype [14]. In order to better understand the origin of the pathogens, in the current study, we sampled the feces of the common food animals during slaughter. Since previously *S*. Tilene has mainly been recovered from African pigmy hedgehogs kept as pets in North America or Europe [15,16], we included hedgehogs, which are common on the grassy pastures in Burkina Faso, in our study.

The specific aims of our study were: first, to estimate the prevalence of *Salmonella* in the feces of slaughtered cattle, poultry and swine, as well as in the feces of hedgehogs in Burkina Faso; second, to identify the serotype of *Salmonella* isolates; third, to determine the sensitivity of the isolates to the antimicrobial agents; and finally, to assess the genetic relatedness of the isolates from the feces of the animals and from the local children using pulsed-field gel electrophoresis (PFGE).

Results

Salmonella prevalence and the serotypes

Salmonella was isolated from 383 (53%) of the total of 729 feces samples from apparently healthy animals. Isolates were obtained from 159 (52%) of the cattle feces, 192 (55%) of the chicken feces, 8 (16%) of the swine feces and 24 (96%) of the hedgehog feces (Table 1). Of the 383 isolates, 382 belonged to *S. enterica* ssp. *enterica* and one was S. *enterica* ssp. *salamae*. 364 of the *S. enterica* ssp. *enterica* isolates could be serotyped in detail, while for 18 isolates only the *Salmonella* group could be assigned. 60 different serotypes were found from the cattle, 41 from the chicken, 5 from the swine and 8 from the hedgehog feces. The predominant serotypes were *S*. Drac and *S*. Muenster in the cattle, *S*. Derby and *S*. Chester in the poultry and *S*. Muenster in both the swine and hedgehog feces. The 3 *S*. Typhimurium isolates from the

5

cattle all belonged to variant Copenhagen. Phage typing divided the *S*. Typhimurium isolates further into three definite phage types: DTs 2, 56 and 116 (Figure 1). In addition, 9 strains were RDNC (reacts but do not conform).

Antimicrobial resistance

On the whole, 52 (14%) of the 383 *Salmonella* isolates were resistant to one or more antimicrobials tested: 23 of these were from the cattle, 23 from the poultry and 6 from the hedgehog feces (Table 1). The salmonella isolates from the swine feces were susceptible to the tested antimicrobials. Six isolates were multiresistant: 4 *S*. Typhimurium isolates from the poultry feces (ampicillin, chloramphenicol, streptomycin, sulfonamides and trimethoprim), 1 *S*. Hato isolate from the poultry feces (ampicillin, streptomycin, sulfonamides, tetracycline, trimethoprim) and 1 *S*. Urbana isolate from the cattle feces (chloramphenicol, trimethoprim, nalidixic acid and mecillinam). Out of the 383 isolates, 247 (64%) showed decreased sensitivity (i.e. were intermediate) to one or more antimicrobial, especially to streptomycin, tetracycline and sulphonamides (Table 1). Two isolates (*S*. Urbana and *S*. Waycross) had decreased sensitivity to ciprofloxacin and one (*S*. Urbana) to cefotaxime. The MIC values for the nalidixic acid resistant isolates were 0.023 (*S*. Munster) and 0.032 (*S*. Urbana).

Genetic relatedness by PFGE

To determine the genotypic relatedness of the *Salmonella* isolates recovered from the cattle, poultry, swine and hedgehog feces and to compare them to human isolates from Burkina Faso, a total of 50 isolates were subjected to PFGE analysis with *Xba*I and *Bln*I restriction enzymes (Figure 1). Genetic relatedness of the isolates belonging to the same serotype ranged from approximately 70% to 100%. *S*. Typhimurium isolates from the poultry and human feces clustered closely together. *S*. Muenster isolates obtained from the cattle and swine feces were different, but both clustered closely together with some hedgehog isolates (Figure 1).

6

Two *S.* Typhimurium var. Copenhagen isolates from the cattle feces clustered together with the *S.* Typhimurium isolates when *Xba*I was used, whereas all three were distinct from S. Typhimurium when *Bln*I was used. *S.* Albany isolates from the cattle and poultry feces clustered separately using both enzymes.

Discussion

We detected high prevalence of *Salmonella enterica* ssp. *enterica* in the feces of the production animals slaughtered for human consumption in Burkina Faso. *Salmonella* was especially common in the poultry (55%) and cattle (52%) feces samples. The levels of *Salmonella* in poultry can vary depending on the country, the nature of the production system and the specific control measures in place. In some EU countries chicken flocks are virtually free from *Salmonella* whereas in the US a contamination rate up to 60% was detected [17]. In Japan, *Salmonella* was isolated from 36% of the broiler fecal samples [18]. In Gambia, the detected rate of *Salmonella* in chicken feces was higher, 67% [19], than what we detected from the chicken feces. In comparison, only 11% of chicken reared at intensive poultry farms in Nigeria were found to be infected [20]. The levels of *Salmonella* rates reported in beef are usually lower than in chicken. *Salmonella* carriage was reported to be 1.4% in cattle in Great Britain [21] and 0.5% in Japan [18]. In Ethiopia, 4% of the feces of slaughtered cattle were contaminated by *Salmonella* [22]. We found 16% of the swine feces samples to be contaminated by *Salmonella*. *Salmonella* contamination rates for pigs reported in literature vary from 9% to 23% in Europe [17,21,23], to 3% of porcine fecal samples in Japan [18] and 8% in Kenya [24]. In accordance to the high rates of *Salmonella* detected in the feces samples, our previous studies on the prevalence of *Salmonella* in retail meats and beef intestines in Burkina Faso also revealed high numbers of *Salmonella*, especially in chicken (37-57%) [13,14].

7

Several of the serotypes isolated in this study, including *S.* Typhimurium, *S.* Muenster, *S.* Derby, *S.* Virchow, *S.* Hato, *S.* Bredeney, *S.* Stanley and *S.* Anatum, have frequently been implicated in outbreaks or sporadic cases of human illness [25]. In Africa, as elsewhere in the world, *S.* Enteritidis and *S.* Typhimurium are the most common causes of human salmonellosis [26]. Interestingly, *S.* Enteritidis was not recovered from the animal feces in our study and not from the human isolates in Burkina Faso either [27]. The main serotypes found in both animal and human feces samples from Burkina Faso included *S.* Typhimurium (from poultry) and *S.* Muenster (from all the studied animal species). *S.* Derby was the most common serotype we detected in the chicken feces, as it was in the chicken carcasses [13,14]. World-wide, a wide range of *Salmonella* serotypes have the ability to colonize poultry: *S.* Typhimurium, *S.* Enteritidis, *S.* Hadar, *S.* Virchow, *S.* Infantis and, recently, *S.* Paratyphi B var. Java have all been frequently isolated from poultry in several countries [17], none of which were among the most common serotypes in poultry in Burkina Faso. Elsewhere in Africa, *S.* Enteritidis was the most common serotype detected in chicken feces in Zimbabwe [28] and *S.* Typhimurium in Algeria [29]. Notably, we isolated one *S.* Typhi strain from the chicken feces, as we did previously from a chicken carcass [14].

The *S.* Typhimurium isolates from chicken feces in Burkina Faso were multi-resistant to the commonly available antimicrobials ampicillin, chloramphenicol, streptomycin, sulfonamides and trimethoprim. This is a typical pattern found in the *Salmonella* strains with a sub-Saharan distinct genotype causing epidemic invasive disease [30]. Bacteremia caused by multi-resistant *S.* Typhimurium strains is a serious public health problem in Africa and they are significantly associated with increased mortality [31]. Such *S.* Typhimurium isolates have been reported from e.g. Zaire [31], Kenya [32], Malawi [32] and Central Africa [33]. Although antimicrobial use for animals is under veterinary prescription control in Burkina Faso, farmers still use unprescribed antimicrobials as growth promoters or treatment for cattle, poultry and swine. This practice leads into a possibility that bacterial resistance developing in the food animals transfers to the

8

human population thus posing a risk for public health by spreading of the resistance [34]. It would be essential to study the genotype of our S. Typhimurium isolates from poultry further in order to know if the invasive genotype also occurs in animals as the environmental reservoirs and host ranges of invasive salmonella strains in Africa are still unknown [35]. Our S. Typhimurium isolates from chicken and humans had the same phage type DT 56. This phage type was in Kenya among the most common phage type from adult patients [36]. In developed countries, a phage type DT 104 has often been associated with outbreaks of multiresistant S. Typhimurium infection in both man and animals [37]. Only two isolates in our study was resistant to the newer antimicrobials; S. Muenster from the poultry feces was resistant to nalidixic acid, as was S. Urbana from the cattle feces, furthermore, its sensitivity to ciprofloxacin and cefotaxime was decreased.

PFGE provides valuable phylogenetic-relationship inference for *Salmonella* at serotype and strain level [38,39]. Our cluster analysis revealed close genetic relationship between some human and animal strains belonging to the same serotypes. Notable similarity of the chicken and human isolates indicates that chicken may be a major source of *Salmonella* transmission to humans. Also in Senegal, a study detected a high degree of similarity among S. Hadar, S. Brancaster and S. Enteritidis from poultry meat and humans by using PFGE [40]. Besides through food, direct transmission from chicken to humans could easily happen in Burkina Faso, since chickens roam free scattering their feces anywhere in the house yards. Although, in these surroundings it is also possible that it is rather chicken which get transiently infected with the typical human *Salmonella* strains. However, the study conducted recently on isolates from infected children and their households in the Gambia did not support the hypothesis that humans and animals living in close contact in the same household carry genotypically similar *Salmonella* serotypes [19].

We found out that the prevalence of *Salmonella* in hedgehog feces was particularly high (96%). In Burkina Faso, hedgehogs live in a variety of habitats where they dig their burrows, spend most of the daylight hours

9

asleep, and emerge at night to forage. Hedgehogs can serve as reservoirs of *Salmonella* in many ways. During the night, villagers go to catch them as a meat source for the next day. During the rainy season, feces of animals including hedgehogs pollute the water sources such as rivers and wells. At the countryside many people are dependent on these sources for their potable water. In developed countries, people having exotic hedgehogs as pets have fallen sick with salmonellosis [10]. In these cases, the commonly detected *Salmonella* serotype has been *S*. Tilene [16]. Since we found several *S*. Tilene strains in our cattle and chicken meat samples during our previous study [13], we wanted to investigate a possible link between the *Salmonella* carriage of the production animals and hedgehogs, which share the same pastures for foraging. Indeed, we found hedgehogs in Burkina Faso to carry many *Salmonella* serotypes common also in the production animals, but no *S*. Tilene was detected, not in feces of the studied hedgehogs or of the other animals.

S. Muenster isolates were obtained from the feces of all the studied animal species and humans and their genetic relatedness in PFGE analysis was 90 to 95%. Thus, it is possible that the same strains of *S*. Muenster are able to infect many different hosts. Hedgehog feces might infect both cattle and swine foraging freely, since *Salmonella* can persist in the environment for several months to more than a year [41,42]. The production animals and the hedgehogs might all be able to transfer *Salmonella* further to the humans. We have previously shown the production animals to be potential carriers of virulent *Escherichia coli* to humans as well [43]. There is no previous information on the frequency of wild animals carrying enteropathogenic bacteria in Burkina Faso, apart from the *Salmonella* carriage of hedgehogs reported here.

Conclusions

10

Our study revealed that both production and some wild animals commonly carry *Salmonella* in Burkina Faso. Some of the isolated *Salmonella* strains were genetically related to the human *Salmonella* strains and resistant to the common antimicrobials. As the humans and animals often live in close vicinity in Africa and the hygiene control of the meat retail chain is defective, high carriage rates of *Salmonella* and other potential pathogens of asymptomatic production animals can pose a major public health problem in Burkina Faso. Therefore, systematic surveillance of the infection sources and routes of the bacterial pathogens especially in the food production chain is needed to control the spread of the pathogens to the consumers.

Methods

Sampling

From 9 March to 25 August 2010, we collected 704 fecal samples from cattle (n=304) and swine (n=50) after slaughter at the central abattoir, and from chickens (n=350) from the local poultry meat sellers in Ouagadougou, Burkina Faso, as previously described [43]. Hedgehogs (n=25) were obtained from different villages across the country. Immediately after the animals were slaughtered, the fecal material was taken aseptically from the large intestine, 1 to 1.5 cm from the rectum. The samples were transported to the laboratory and kept at 4°C until the microbiological examination was started within 8 hours.

Salmonella isolation and phenotyping

From each fecal sample, 25 g was enriched in 225 ml of buffered peptone water (Liofilchem, Teramo, Italy) at 37°C for 24 h. After that, 0.1ml of an enriched sample was transferred into 10 ml of Rappaport-Vassiliadis broth and incubated at 42°C for additional 24 h before plating a loopful on Xylose Lysine

11

Deoxycholate (XLD) agar (Oxoid, Basingstoke, England). Identity of the colonies with black center were confirmed biochemically using lysine and triple sugar iron agars and with API 20E (Biomerieux, Marcy l'Etoile, France). *Salmonella* isolates were serotyped with the somatic O and flagellar H anti-sera according to the Kauffman-White scheme [44]. Isolates of serotypes Typhimurium (including var. Copenhagen) were further phage typed [45].

Antimicrobial susceptibility testing

Antimicrobial susceptibility of the isolates was tested by a standard disk diffusion method, and *Escherichia coli* RHE 6715 (ATCC 25922) was used for validating the antimicrobial test results [46]. The antimicrobial agents used were ampicillin (10 µg), chloramphenicol (30 µg), streptomycin (10 µg), sulphonamides(3 µg), trimethoprim (5 µg), tetracycline (30 µg), gentamicin (10 µg), nalidixic acid (30 µg), ciprofloxacin (5 µg), cefotaxime (30 µg), mecillinam (10 µg), imipenem (10 µg). Minimal inhibitory concentration (MIC) for ciprofloxacin was determined by E-test (AB Biodisk, Solna Sweden) to the isolates resistant to nalidixic acid. MIC breakpoint $\leq$ 1mg/l was interpreted as susceptible [46].

Genotyping

For Pulsed-Field Gel Electrophoresis (PFGE) the PulseNet protocol for *Salmonella* was used with the *Xba*I and *Bln*I restriction enzymes [47]. Briefly, agarose-embedded DNA was digested with 15 U of restriction enzyme (Fermentas International, Burlington, Ontario) at 37°C overnight. The restriction fragments were separated by electrophoresis in 0.5x TBE (HEPES for *S*.Ouakam) running buffer at 14°C for 20 h using the CHEF Mapper electrophoresis system (Bio-Rad Laboratories, Hercules, California, USA) with pulse times of 2 to 63 s, 120° angle, and 6.0 V/cm gradient. The agarose gels were stained with ethidium bromide, and the DNA banding patterns were analyzed by BioNumerics 5.10. *Salmonella* Braenderup H9812 was used as a standard. The 19 human *Salmonella* isolates included in the PFGE analysis were obtained from the

12

National Public Health Laboratory in Ouagadougou, Burkina Faso and described in [4]. They included serotypes Typhimurium (13 strains), Virchow (2), Ouakam (2) and Muenster (2).

Ethical considerations

Permission to conduct this study was obtained from the slaughterhouse authorities and the study protocol was approved by the Ethical Committee of Burkina Faso.

Competing interests

The authors declare that they have no competing interests.

Authors' contributions

AK carried out the sampling and strain characterization and drafted the manuscript, TL and LA participated in the PFGE analysis, AST and NB supervised the sampling and strain isolation, AS and KH supervised the strain characterization and participated in writing the manuscript. All authors read, commented on and approved of the final manuscript.

Acknowledgements

This study was funded by the Academy of Finland grant 122600 to collaboration between the Finnish National Institute for Health and Welfare (THL) and CRSBAN/University of Ouagadougou and CRSBAN/University of Ouagadougou and by the International Foundation for Science (IFS) grant to AK. We thank the personal from the national slaughterhouse of Ouagadougou and the poultry sellers for the good collaboration. We also thank the personnel of the Bacteriology Unit at THL for their assistance in sero- and phage typing.

13

References

1. Majowicz SE, Musto J, Scallan E, Angulo FJ, Kirk M, O'Brien SJ, Jones TF, Fazil A, Hoekstra RM: **The Global Burden of Nontyphoidal *Salmonella* Gastroenteritis.** *Clin Infect Dis* 2010, **50**:882–889.

2. Bryce J, Boschi-Pinto C, Shibuya K, Back RE: **WHO estimates of the causes of death in children.** *Lancet* 2005, **365**:1147–1152.

3. Morpeth SC, Ramadhani HO, Crump JA: **Invasive non-Typhi *Salmonella* disease in Africa.** *Clin Infect Dis* 2009, **49**:606-611.

4. Acha PN, Szyfres B: **Salmonellosis.** In *Zoonoses and Communicable Diseases Common to Man and Animals, Volume I: Bacterioses and Mycoses.* 3rd edition. Edited by Acha PN, Szyfres B. Pan American Health Organization, Washington, D.C. 2001: 233-246.

5. Kariuki S, Revathi G, Kariuki N, Kiiru J, Mwituria J, Muyodi J, Githinji JW, Kagendo D, Munyalo A, Hart CA: **Invasive multidrug-resistant non-typhoidal *Salmonella* infections in Africa: zoonotic or anthroponotic transmission?** *J Med Microbiol* 2006, **55**:585-591.

6. Thorns CJ: **Bacterial food-borne zoonoses.** *Rev Sci Tech* 2000, **19**:226–239.

7. Gillespie IA, O'Brien SJ, Adak GK, Ward LR, and Smith HR: **Foodborne general outbreaks of *Salmonella* Enteritidis phage type 4 infection, England and Wales, 1992–2002: where are the risks?** *Epidemiol Infect* 2005, **133**:795-801.

8. Stopforth JD, Lopes M, Shultz JE, Miksch RR, Samadpour M: **Location of bung bagging during beef slaughter influences the potential for spreading pathogen contamination on beef carcasses.** *J Food Prot* 2006, **69**:1452-1455.

14

9. Glaser CA, Angulo FJ, Rooney J: **Animal-associated opportunistic infections in HIV-infected persons.** *Clin Infect Dis* 1994, **18**:14-24.

10. Riley PY, Chomel BB: **Hedgehog zoonoses.** *Emerg Infect Dis* 2005, **11**:1-5.

11. White DG, Zhao S, Sudler R, Ayers S, Friedman S, Chen S, McDermott PF, McDermott S, Wagner DD, Meng J: **The isolation of antibiotic resistant *Salmonella* from retail ground meat.** *New Engl J Med* 2001, **345**:1147–1154.

12. Threlfall EJ: **Antimicrobial drug resistance in *Salmonella*: problems and perspectives in food- and water-borne infections.** *FEMS Microbiol Rev* 2002, **26**:141–148.

13. Kagambèga A, Haukka K, Siitonen A, Traoré AS, Barro N: **Prevalence of *Salmonella enterica* and the hygienic indicator *Escherichia coli* in raw meat at markets in Ouagadougou, Burkina Faso.** *J Food Prot* 2011, **74**:1547-1551.

14. Kagambega A, Barro N, Traoré AS, Siitonen A, Haukka K: **Characterization of *Salmonella enterica* and detection of the virulence genes specific to diarrheagenic *Escherichia coli* from poultry carcasses in Ouagadougou, Burkina Faso.** *Foodborne Pathog Dis* 2012, **9**:589-593.

15. CDC: **African pygmy hedgehog-associated salmonellosis.** *MMWR Morb Mortal Wkly Rep* 1995, **44**:462-463.

16. Craig C, Styliadis S, Woodward D, Werker D: **African pygmy hedgehog-associated *Salmonella tilene* in Canada.** *Can Commun Dis Rep* 1997, **23**:129-131.

17. Mølbak K, Olsen JE, Wegener HC: *Salmonella* **infections.** In *Foodborne Infections and Intoxications*. 3rd edition. Edited by Riemann HP, Cliver DO. Elsevier, The Netherlands. 2006: 57-136.

15

18. Ishihara K, Takahashi T, Morioka A, Kojima A, Kijima, Asai T, Tamura Y: **National surveillance of** *Salmonella enterica* **in food-producing animals in Japan.** *Acta Vet Scand* 2009, **51**:35.

19. Dione MM, Ikumapayi UN, Saha D, Mohammed NI, Geerts S, Ieven M, Adegbola RA, Antonio M: **Clonal differences between non-typhoidal** *salmonella* **(NTS) recovered from children and animals living in close contact in the Gambia.** *PLoS Negl Trop Dis* 2011, **5**:1148.

20. Fashae K, Ogunsola F, Aarestrup FM, Hendriksen RS: **Antimicrobial susceptibility and serovars of** *Salmonella* **from chickens and humans in Ibadan, Nigeria.** *J Infect Dev Ctries* 2010, **4**:484-494.

21. Milnes AS, Sayers AR, Stewart I, Clifton-Hadley FA, Davies RH, Newell DG, Cook AJ, Evans SJ, Smith RP, Paiba GA: **Factors related to the carriage of Verocytotoxigenic** *E. coli, Salmonella,* **thermophilic** *Campylobacter* **and** *Yersinia enterocolitica* **in cattle, sheep and pigs at slaughter.** *Epidemiol Infect* 2009, **137**:1135-1148.

22. Molla B, Alemayehu D, Salah W: **Sources and distribution of** *Salmonella* **serotypes isolated from food animals, slaughterhouse personnel and retail meat products in Ethiopia: 1997-2002.** *Ethip J Health Dev* 2003, **17**:63-70.

23. Lomonaco S, Decastelli L, Bianchi DM, Nucera D, Grassi MA, Sperone V, Civera T: **Detection of** *Salmonella* **in finishing pigs on farm and at slaughter in Piedmont, Italy.** *Zoonoses Public Health* 2009, **56**:137-144.

24. Kikuvi GM, Ombui JN, Mitema ES: **Serotypes and antimicrobial resistance profiles of** *Salmonella* **isolates from pigs at slaughter in Kenya.** *J Infect Dev Ctries* 2010, **4**:243-248.

16

25. CDC: *Salmonella* surveillance: annual summary, 2006. Centers for Disease Control and Prevention, Atlanta, GA. 2008. http://www.cdc.gov/ncidod/dbmd/phlisdata/salmonella.htm.

26. Hendriksen RS, Vieira AR, Karlsmose S, Lo Fo Wong DM, Jensen AB, Wegener HC, Aarestrup FM: **Global monitoring of *Salmonella* serovar distribution from the World Health Organization Global Foodborne Infections Network Country Data Bank: results of quality assured laboratories from 2001 to 2007.** *Foodborne Pathog Dis* 2011, 8:887-900.

27. Bonkoungou IJO, Haukka K, Österblad M, Hakanen AJ, Traoré AS, Barro N, Siitonen A: **Bacterial and viral etiology of childhood diarrhea in Ouagadougou, Burkina Faso.** *BMC Pediatr* 2013, 13:36.

28. Makaya PV, Matope G, Pfukenyi DM: **Distribution of *Salmonella* serovars and antimicrobial susceptibility of *Salmonella* Enteritidis from poultry in Zimbabwe.** *Avian Pathology* 2012, 41:221-226.

29. Ayachi A, Alloui N, Bennoune O, Kassah-Laouar A: **Survey of *Salmonella* serovars in broilers and laying breeding reproducers in Eastern Algeria.** *J Infect Dev Ctries* 2010, 4:103-106.

30. Kingsley RA, Msefula CL, Thomson NR, Kariuki S, Holt KE, Gordon MA, Harris D, Clarke L, Whitehead S, Sangal V, Marsh K, Achtman M, Molyneux ME, Cormican M, Parkhill J, MacLennan CA, Heyderman RS, Dougan G: **Epidemic multiple drug resistant *Salmonella* Typhimurium causing invasive disease in sub-Saharan Africa have a distinct genotype.** *Genome Res* 2009, 19:2279-2287.

31. Green SDR, Cheesbrough JS: ***Salmonella* bacteraemia among young children at a rural hospital in western Zaire.** *Ann Trop Paediatr* 1993, 13:45-53.

17

32. Leegaard TM, Van Gestel MH, Petit PLC, Van de Klundert JAM: **Antibiotic resistance mechanisms in *Salmonella* species causing bacteraemia in Malawi and Kenya.** *APMZS* 1996, **104**:302-306.

33. Lepage P, Bogaerts J, Nsengumuremyi F, Hitimana DG, Van Goethem C, Vandepitte J, Butzler JP: **Severe multiresistant *Salmonella typhimurium* systemic infections in Central Africa - clinical features and treatment in a paediatric department.** *J Antimicrob Chemother* 1984, **14** (Suppl B): 153-159.

34. Ungemach FR, Müller-Bahrdt D, Abraham G: **Guidelines for prudent use of antimicrobials and their implications on antibiotic usage in veterinary medicine.** *Inter J Med Microbiol* 2006, **296**:33–38.

35. Feasey NA, Dougan G, Kingsley RA, Heyderman RS, Gordon MA: **Invasive non-typhoidal *Salmonella* disease: an emerging and neglected tropical disease in Africa.** *Lancet* 2012, **379**:2489–2499.

36. Kariuki S, Gilks C, Kimari J, Muyodi J, Waiyaki P, Hart CA: **Analysis of *Salmonella enterica* serotype Typhimurium by phagetyping, antimicrobial susceptibility and pulsed-field gel electrophoresis.** *J Med Microbiol* 1999, **48**:1037-1042.

37. Threlfall EJ: **Epidemic *Salmonella typhimurium* DT104 – a truly international multiresistant clone.** *J Antimicrob Chemother* 2000, **46**:7-10.

38. Harbottle H, White DG, McDermott PF, Walker RD, Zhao S: **Comparison of multilocus sequence typing, pulsed-field gel electrophoresis, and antimicrobial susceptibility typing for**

characterization of *Salmonella enterica* serotype Newport isolates. *J Clin Microbiol* 2006, **44**:2449–2457.

39. Gaul SB, Wedel S, Erdman MM, Harris DL, Harris IT, Ferris KE, Hoffman L: **Use of pulsed-field gel electrophoresis of conserved *XbaI* fragments for identification of swine *Salmonella* serotypes.** *J Clin Microbiol* 2007, **45**:472–476.

40. Cardinale E, Perrier Gros-Claude JD, Rivoal K, Rose V, Tall F, Mead GC, Salvat G: **Epidemiological analysis of *Salmonella enterica* ssp. *enterica* serovars Hadar, Brancaster and Enteritidis from humans and broiler chickens in Senegal using pulsed-field gel electrophoresis and antibiotic susceptibility.** *J Appl Microbiol* 2005, **99**:968-977.

41. Winfield MD, Groisman EA: **Role of nonhost environments in the lifestyles of *Salmonella* and *Escherichia coli.*** *Appl Environ Microbiol* 2003, **69**:3687–3694.

42. Parker CT, Huynh S, Quinones B, Harris LJ, Mandrell RE: **Comparison of genotypes of *Salmonella enterica* serovar Enteritidis phage type 30 and 9c strains isolated during three outbreaks associated with raw almonds.** *Appl Environ Microbiol* 2010, **76**:3723–3731.

43. Kagambèga A, Martikainen O, Siitonen A, Traoré AS, Barro N, Haukka K: **Prevalence of diarrheagenic *Escherichia coli* virulence genes in the feces of slaughtered cattle, chickens, and pigs in Burkina Faso.** *MicrobiologyOpen* 2012; 1:276-284.

44. Popoff MY, Bockemuhl J, Gheesling LL: **Supplement 2002 (no. 46) to the Kauffmann-White scheme.** *Res Microbiol* 2004, **155**:568-570.

45. Anderson ES, Ward LR, Saxe MJ, de Sa JD: **Bacteriophage typing designations of *Salmonella typhimurium.*** *J Hyg* (Lond) 1977, **78**:297–300.

19

46. CLSI (Clinical and Laboratory Standards Institute): **Methods for Dilution Antimicrobial Susceptibility Tests for Bacteria that Grow Aerobically.** 2009. http://www.clsi.org/source/orders/free/m07-a8.pdf. Accessed 1. Dec 2011.

47. PulseNet: **One-day (24–48 h) standardized laboratory protocol for molecular subtyping of** *Escherichia coli* **O157:H7, non-typhoidal** *Salmonella* **serotypes, and** *Shigella sonnei* **by pulsed field gel electrophoresis (PFGE).** 2002. http://www.cdc.gov/pulsenet/protocols/ecoli-salmonella-shigella-protocols.pdf. Accessed 11 Jul 2006.

20

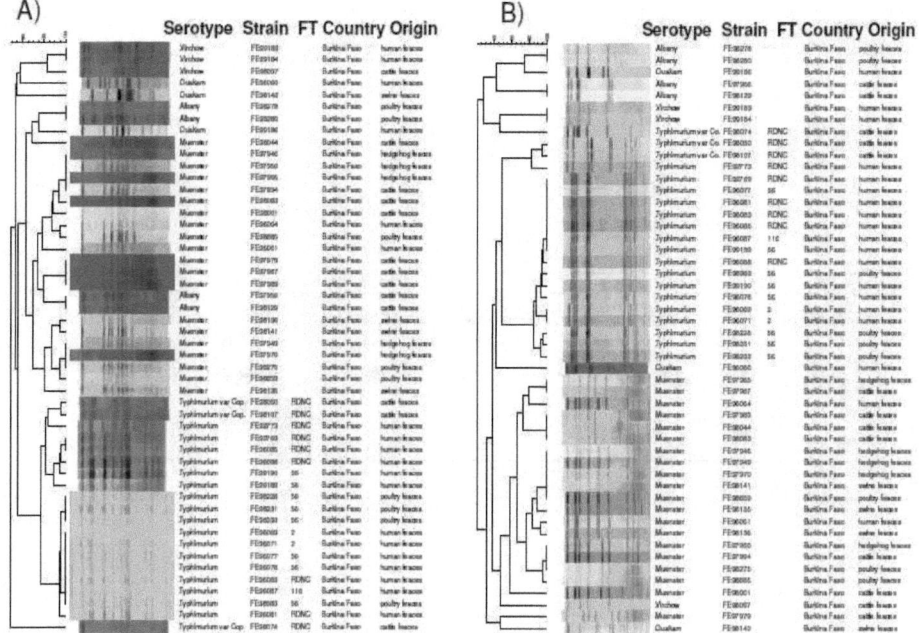

Figure 1:

- Cluster analysis of five different salmonella serotypes (Muenster n=20, Typhimurium n=17, Albany n=4, Typhimurium var Copenhagen n=3 and Ouakam n=1) which were isolated from both human and animal feaces. **A)** *Xba*l-PFGE analysis and **B)** *Bln*l-PFGE analysis.

Table 1: *Salmonella enterica* serotypes isolated from cattle, poultry, swine and hedgehog feces and their antimicrobial resistance patterns.

Salmonella serotypes	Cattle feces (n=304)	Poultry feces (n= 350)	Swine feces (n=50)	Hedgehog feces (n= 25)	Total (n=729)	Antimicrobial resistance patterns Resistant [a]	Intermediate [a]
S. Abaetetuba	1	1	-	-	2	-	1Pstr-tet, 1Cstr
S. Abony	-	1	-	-	1	-	-
S. Adelaide	-	1	-	-	1	-	-
S. Agona	-	3	-	-	3	-	1Pstr-sul, 1Cstr
S. Albany	2	2	-	-	4	-	1Ptet, 1Cstr
S. Anatum	-	1	-	-	1	-	1Pstr
S. Ank	-	1	-	4	5	-	4Hstr, 1Pstr
S. Antwepen	1	-	-	-	1	-	1Cstr
S. Apeyeme	2	3	-	-	5	2Cstr	3Pstr

S. Banana	1	2	-	1	4	1Hstr	1Cstr
S. Bareilly	1	-	-	-	1	-	1Cstr
S. Bargny	1	-	-	-	1	-	1Cstr
S. Binningen	-	2	-	-	2	-	-
S. Brancaster	1	3	-	-	4	-	1Cstr, 1Pstr, 1Pstr-tet
S. Bredeney	5	2	-	-	7	-	4Cstr, 1Pstr
S. Brive	1	-	-	-	1	-	1Cstr
S. Carmel	1	-	-	-	1	-	-
S. Carno	1	-	-	-	1	-	-
S. Chandans	2	-	-	-	2	-	2Cstr
S. Chester	1	31	-	-	32	1Pmec	29Pstr, 1Cstr, 1Pstr-tet
S. Chomedey	4	-	-	-	4	-	4Cstr
S. Colindale	1	-	-	-	1	-	1Cstr
S. Colobane	2	-	-	-	2	1Cstr	1Cstr
S. Dahra	2	-	-	-	2	-	1Cstr-tet
S. Dakar	1	-	-	-	1	1Cstr	-
S. Derby	-	51	-	-	51	5Ptet, 3Pstr, 1Pchl, 1Psul	22Pstr , 1Psul, 1Psul-tet, 7Pstr-tet, 7Pstr-sul, 2Pstr-sul-tet

23

S. Drac	26	-	-	1	27	4Cstr	1Hstr, 22Cstr
S. Duisburg	-	1	-	-	1	-	1Pstr
S. Eastbourne	2	2	-	-	4	-	2Cstr, 1Pstr, 1Pstr-tet
S. Farakan	3	-	-	-	3	1Cstr	1Cstr
S. Freetown	-	1	-	-	1	-	1Pstr
S. Fresno	-	4	-	-	4	1Pstr	1Pstr
S. Frintrop	1	-	-	-	1	-	1Cstr
S. Fufu	1	-	-	-	1	-	1Cstr
S. Galiema	-	2	-	-	2	-	2Pstr
S. Gokul	1	-	-	-	1		1Cstr
S. Hato	5	22	-	-	27	1Pamp-str-sul-tet-tmp, 1Pamp, 1Pstr	8Pstr, 1Psul-tet, 2Pstr-tet, 1Ptet, 1Cstr
S. Hillingdon	-	1	-	-	1	-	1Pstr
S. Ikeja	1	-	-	-	1	-	1Cstr
S. Ilala	2	-	1	-	2	-	1Sstr
S. Kaapstad	-	4	1	-	5	-	1Pstr, 1Sstr
S. Kalamu	1	-	-	-	1	-	-

24

S. Kalina	2	-	-	-	2	-	1Cstr
S. Kingston	2	3	-	-	5	-	1Pstr, 1Cstr
S. Kokomlemle	2	1	-	-	3	-	1Pstr, 1Cstr
S. Korlebu	2	-	-	-	2	2Cstr	-
S. Lagos	4	2	-	-	6	2Pstr	1Ptet, 2Cstr
S. Moero	1	-	-	-	1	-	-
S. Monschaui	1	1	-	3	5	3Hstr	1Pstr
S. Muenster	17	6	3	11	37	1Camp, 1Cstr, 1Pnal, 1Hsul, 1Hstr	5Hstr, 6Cstr, 4Pstr, 2Sstr, 1Htet
S. Nima	3	-	-	-	3	-	-
S. Nottingham	2	1	-	-	3	-	1Pstr-tet
S. Oranienburg	1	-	-	-	1	-	1Cstr
S. Othmarschen	1	-	-	-	1	1Cstr	-
S. Ouakam	-	-	1	-	1	-	1Sstr
S. Poona	2	1	-	-	3	-	1Pstr, 2Cstr
S. Rissen	1	-	-	-	1	-	-
S. Ruiru	8	-	-	-	8	1Cstr, 1Cstr-tet	3Cstr

S. Saintpaul	-	1	-	-	1	-	1Ptet
S. Salford	1	-	-	-	1	-	-
S. Schwarzengrund	1	3	-	-	4	-	1Cstr , 3Pstr
S. Senftenberg	-	8	-	2	10	-	4Pstr, 2Pstr-tet, 1Pstr-sul-tet
S. Shangani	-	1	-	-	1	-	1Pstr -sul
S. Soumbedioune	4	-	-	-	4	-	3Cstr
S. Stanley	-	-	-	1	1	-	1Hstr
S. Stanleyville	-	1	-	-	1	-	1Pstr-tet
S.Tennessee	3	-	-	-	3	-	1Cstr
S. Trachau	1	1	-	-	2	1Cstr	1Pstr
S. Typhi	-	1	-	-	1	1Pstr	-
S. Typhimurium	3	4	-	-	7	4Pamp-chl-str-sul-tmp, 3Cstr	-
S. Umbadah	1	-	-	-	1	-	-
S. Umbilo	1	-	-	-	1	-	1Cstr
S. Urbana	13	1	2	-	16	1Cchl-tmp-nal-mec	4Cstr, 1Cstr-ftx, 2Cstr-tet, 1Cstr-cip, 1Pstr, 1Sstr
S. Virchow	1	-	-	-	1	-	1Cstr

26

S. Waycross	2	1	-	-	3	1Cstr	1Cstr, 1Pcip
S. Yoruba	1	-	-	-	1	-	1Cstr
S. group B 4,5,12:-:-	1		-	-	1	1Cstr-tet	-
S. group C 6,7,14:d:-	1	9	-	-	10	-	5Pstr-sul, 4Pstr, 1Cstr
S. group E 3,10:e,h:-	1	5	-	-	6	-	1Pstr-sul-tet, 1Pstr, 1Cstr
S. group G 13,22:z:-	-	-	-	1	1	-	1Hstr
Salmonella enterica ssp. *salamae*	1	-	-	-	1	-	-
Total	159	192	8	24	383	52	247
	(52%)	(55%)	(16%)	(96%)	(53%)	(7%)	(34%)

[a] For example, entry 7Pstr-tet, means that 7 isolates from poultry feces were resistant/intermediate to streptomycin and tetracycline. Abbreviations: **C**, cattle feces; **P**, poultry feces; **S**, swine feces; **H**, hedgehog feces, amp, ampicillin; chl, chloramphenicol; str, streptomycin; sul, sulphonamides; tmp, trimethoprim; tet, tetracycline; nal, nalidixic acid; cip, ciprofloxacin; ftx, cefotaxime; mec, mecillinam.

27

Chapitre 4: Discussion générale

Les maladies transmises par les aliments posent un problème majeur de santé publique à travers le monde, et plus particulièrement dans les pays en développement où l'hygiène alimentaire reste un défi à relever (Whyte *et al.*, 1997). Au Burkina Faso, les agents principaux en cause dans ces maladies sont les bactéries surtout les Salmonelles et les pathovars de *Escherichia coli* (Barro *et al.*, 2005). Ces pathogènes sont d'origine zoonotique et se transmettent le plus souvent par des aliments d'origine animale (Blanco *et al.*, 2004; Kagambèga *et al.*, 2011). Pourtant, très peu de données scientifiques existent sur la prévalence de *Salmonella* ou pathovars de *E. coli* sur des viandes ou produits carnés au Burkina Faso.

Pour apporter plus d'informations sur ces pathogènes circulant au Burkina Faso, nos premiers travaux ont porté essentiellement sur la prévalence des *Salmonella* et *E. coli* dans les viandes de bœuf, de mouton et de poulet. L'enquête sur les conditions sociodémographiques et professionnelles des bouchers lors de ces travaux, nous a révélé le faible niveau d'instruction, le manque de formation en bonnes pratiques d'hygiène et le manque de précautions sanitaires de manipulation des viandes qui sont des facteurs majeurs de contamination. Cela a été démontré par plusieurs auteurs (Umoh et Odoba, 1999; King *et al.*, 2000; Barro *et al.*, 2002). La présence d'animaux errants et d'insectes sur les lieux de vente sont des sources importantes de contamination, également signalée par d'autres études (Umoh et Odoba, 1999; Barro *et al.*, 2008).

Les résultats d'analyses microbiologiques montrent des prévalences de *Salmonella* allant de 13% sur les viandes de boeuf à 57% sur les carcasses de poulet. Plusieurs auteurs ont montré que les viandes de poulet constituent une source importante de salmonellose humaine (Cardinale *et al.*, 2004; Wilson, 2002). Au sénégal, l'étude menée par Stevens *et al.* (2006) a révélé une prévalence de *Salmonella* de 90% sur la viande de bœuf, ce qui est très élevé comparé à 13% rapportés dans notre étude. Par contre, en Irlande, une prévalence similaire a celle rapportée dans cette étude a été détectée sur des viandes de bœuf (McEvoy *et al.*,

2003). La prévalence élevée de *Salmonella* sur les viandes dans cette étude, peut être expliquée par le portage asymptomatique de ces pathogènes par les animaux qui peuvent contaminer les carcasses lors de l'abattage mais aussi par les mauvaises conditions d'hygiène lors de la vente démontrées au cours de l'enquête.

Le sérotypage des *Salmonella* nous a révélé 17 sérotypes différents de *Salmonella enterica* sur les viandes. Ces sérotypes ont été déjà isolés chez des patients malades par plusieurs auteurs (Leegaard *et al.*, 1996; Herrero *et al.*, 2006; Hald *et al.*, 2007; Bonkoungou *et al.*, 2011). Ce qui montre que ces viandes peuvent être source de salmonellose humaine si des règles d'hygiène ne sont pas respectées lors de la prépartion des viandes afin d'éviter toute contamination croisée.

La diversité de sérotypes sur des échantillons de viande a été aussi démontrée dans d'autre pays d'Afrique de l'Ouest par plusieurs auteurs (Dione *et al.*, 2011; Raufu *et al.*, 2009; Fashae *et al.*, 2010; Nzouankeu *et al.*, 2010).

Le sérotype Typhi a été détecté sur des carcasses de poulet lors de cette étude. La présence de cette bactérie sur les carcasses est d'une grande importance car elle est responsable de la fièvre typhoïde chez l'homme qui est toujours un réel problème de santé dans les pays en développement (Kariuki, 2008). En plus, ce sérotype est typique à l'homme et sa présence sur les carcasses explique sans doute des contaminations croisées dues aux mauvaises pratiques d'hygiène lors de l'abattage et/ou de la vente.

Parmi les pathovars de *E. coli*, les STEC ont été les plus nombreux sur les viandes de bœuf (43%) et mouton (20%) et rarement isolés sur les poulets. Ces prévalences sont plus élevées par rapport à celle (4%) rapportée au Nigeria sur des échantillons de viande (Ojo *et al.*, 2010). Partout dans le monde, les infections à STEC sont le plus souvent liées à la consommation de viande de bœuf (Hussein, 2007). Aussi, la capacité des STEC à induire une infection à faible dose comparativement à EPEC, ETEC et EAEC montre une importance de sa surveillance au niveau des aliments. EPEC, ETEC et EAEC ont été détectés à de faibles proportions par rapport à STEC sur les viandes à l'exception du poulet qui semble être la principale source de EPEC. Des résultats similaires ont été obtenus en Inde par

Farooq *et al.* (2009). La PCR nous a révélé différente proportion au niveau des gènes de virulence de STEC: 32% des STEC détectés sur les viandes possédaient les gènes responsables de la production de shiga-toxine 1 et 2 (*stx*1 et *stx*2), 22% possédaient seulement *stx*1 et 6% posssédaient seulement *stx*2. Contrairement à Lee *et al.* (2009) qui a observé que 64% des STEC isolés de la viande possédaient *stx*2, 14% possédaient *stx*1 et *stx*2, 14% possédaient *stx*2 et *eae*A, et 7% possédaient *stx*1 et *eae*A.

La présence de plus d'un pathovar dans un même échantillon de viande a été détectée dans beaucoup de nos échantillons. Cette situation peut favoriser le développement de bactéries hybrides par un transfert de gène entre différents pathovars. Recemment, la bactérie impliquée dans la grande épidémie en Allemagne resultait d'une combinaison entre STEC et EAEC (EFSA, 2011).

Cette première partie nous a permis de savoir le rôle que peut jouer la viande dans les infections bactériennes au Burkina Faso, vue les différentes prévalences de *Salmonella* et des pathovars de *E. coli*. Mais cette partie nous amène aussi à se poser de questions sur les réservoirs de ces pathogènes. Quelle peut être la source de distribution de ces pathogènes sur les viandes en particulier et dans l'environnement de facon générale?

Salmonella et *E. coli* font partie de la flore intestinale des animaux. Ainsi nous avons mené pour la première fois au Burkina Faso, une étude sur la prévalence de ces entéropathogènes au niveau des animaux domestiques tels que le bœuf, le poulet et le porc et d'animaux sauvages comme le hérisson.

La prévalence de *Salmonella enterica* a été de 52% chez les bovins, 55% chez le poulet, 16% chez le porc et 96% chez le hérisson. De faible proportion de *Salmonella* ont été rapportées dans des fèces de bovins (4,2%) en Ethiopie (Molla *et al.*, 2003) et de porc (8,6%) au kenya (Kikuvi, 2010). Par contre 67% de fèces de poulet ont été contaminés par *Salmonella* en Gambie (Dione *et al.*, 2011).

Nos résultats montrent également que le hérisson est une source de Salmonelloses humaines car des sérotypes communs ont été identifiés chez le hérisson, chez certains animaux et des souches provenant de patients malades du

Burkina Faso. Ces résultats sont en concordance avec ceux d'autres auteurs qui ont montrés que le hérisson d'Afrique a été déjà associé à des cas de Salmonellose humaine chez des personnes qui l'avaient comme animal de compagnie (Lispky et Tannio, 1995; Woodward *et al.*, 1997).

Au total, 75 sérotypes différents de *Salmonella enterica* ont été identifiés sur les animaux englobant ceux précédemment isolés des viandes à l'exception de Tilene et Typhi. Cette situation montre clairement la circulation de *Salmonella* entre animaux, viande et homme.

Les salmonelles isolées au niveau de la viande, d'animaux et chez l'homme ont été résistantantes aux antibiotiques les plus accessibles (Ampicilline, Chloramphenicole, streptomycine, sulfonamides et trimethoprime). Cela pourrait s'expliquer par l'utilisation anarchique de ces antibiotiques aussi bien pour le traitement des hommes que pour le traitement des animaux malades, et pourtant, toute utilisation d'antibiotique favorise la sélection de bactéries résistantes. Il serait donc important de mettre en place un système de surveillance de l'utilisation de ces antibiotiques au niveau clinique et vétérinaire.

Les résultats du PFGE ont révélé des salmonelles isolées de différentes sources et qui présentaient des taux de similitude très élevés. Par exemple S. Typhimurium isolées des fèces de poulet et des patients malades était très similaires (90-95%); en plus ils étaient multi-résistants et avaient le même profil de résistance. Ce qui explique la propagation des pathogènes multirésistants entre les animaux et l'homme, comme déjà observée par Ungemach *et al*. (2006). Cette situation est très inquiétante au Burkina Faso, vue la promiscuité entre l'homme et les animaux domestiques.

En ce qui concerne les pathovars de *E. coli*, ils ont été détectés à de grande proportion au niveau des fèces comparativement aux viandes et en plus tous les 5 pathovars étaient présents. Cela explique le rôle de réservoir de ces animaux domestiques qui sont utilisés comme source de viande. La présence de ces pathogènes sur les viandes au préalable explique sans doute des contaminations croisées lors de l'abattage. Les STEC ont été les plus dominants chez le bœuf par contre les EPEC ont été prédominants chez les poulets. Ce qui est en concordance avec des études antérieures qui ont soulignés que les bovins sont des principaux réservoirs de STEC et les poulets des réservoirs de EPEC (Caprioli *et al.*, 2005; Cookson *et al.*, 2006; Lee *et al.*, 2009). Mais il ressort aussi de notre étude que les porcs sont probablement des réservoirs de STEC, EPEC et EAEC, ce qui mérite une attention particulière car les viandes de porcs sont les moins chères partout dans le monde.

Ces pathovars ont été également isolés de patients malades au Burkina Faso par Bonkoungou *et al.* (2011), traduisant sans doute la contamination des hommes par les animaux à travers des aliments ou par contact direct.

Le taux élevé des salmonelles et pathovars de *E. coli* au niveau des animaux montre leur rôle de réservoirs de ces pathogènes. Pourtant, ces animaux domestiques en plus de la contamination des viandes sont également source de contamination fécale de l'eau de boisson et des légumes. Ce qui favorise un transfert direct des pathogènes à l'homme. Des mesures d'hygiène doivent être prises afin de réduire la propagation de ces bactéries zoonotiques entre animaux et homme.

CONCLUSION GENERALE ET PERSPECTIVES

Conclusion générale et perspectives

L'étude que nous avons menée sur la caractérisation phénotypique et moléculaire de *Salmonella* et de *Escherichia coli* nous a permi de comprendre leur principale source et mode de contamination en rapport avec les systèmes d'élevages, le processus de production des carcasses et les conditions de vente.

Elle porte, à notre connaissance les premières données sur les contaminations de la filière viande par les salmonelles et pathovars de *E. coli*, avec un taux de prévalence relativement élevé aussi bien au niveau des viandes qu'au niveau des animaux. Cette situation est due aux conditions précaires d'élevage et d'abattage, mais aussi aux mauvaises pratiques d'hygiène lors de l'exposition des viandes au Burkina Faso. Elle nous renseigne sur les sérotypes les plus fréquemment isolés, à savoir *S*. Derby, *S*. Muenster, *S*. Virchow, *S*. Infantis, *S*. Albany et *S*. Typhimurium.

Plusieurs profils de résistance aux antibiotiques des salmonelles ont été mis en évidence, suggérant des liens épidémiologiques à différents niveaux: environnemental et clinique. Les profils de résistance aux antibiotiques nous ont permis d'envisager certaines hypothèses épidémiologiques, mais uniquement en confortant la caractérisation génotypique des souches par l'électrophorèse en champ pulsé (PFGE, Pulsotypage). Cette méthode qui s'est confirmé très discriminant, nous a informé sur les liens génétiques et par conséquent nous a suggéré des liens épidémiologiques. La similitude entre les sérotypes communs isolés de même et/ou différents sources a été démontrée, confirmant nos hypothèses de diffusion ou de persistance de certaines souches de *Salmonella* entre les élevages, les abattoirs et les consommateurs. Des études plus poussées sur les différentes sources nous aideraient sûrement à mieux connaître le cycle de diffusion des salmonelles au Burkina Faso et à mieux les combattre.

La similitude des pulsotypes parmi les isolats humains et animaux, interpelle à la nécessité d'une surveillance pour le respect des normes d'élevage et d'abattage, particulièrement les mesures d'hygiène individuelle et collective mais aussi le contrôle de tous les intrants. En plus, comme la majorité des souches de salmonelles résistantes aux antibiotiques sont toujours associées à des infections plus graves (septicémiques) et à un taux d'hospitalisations plus élevé, il serait capital d'accentuer la surveillance de l'utilisation des antibiotiques, notamment en médecine vétérinaire, dans le but de prévenir l'augmentation des résistances aux molécules récentes. Il est aussi impératif de songer à l'association de techniques d'analyses phénotypiques et génotypiques efficaces, qui se sont avérées complémentaires, pour tracer avec précision la diffusion ou la persistance de ces sérotypes, dans une région donnée, et particulièrement les multi-résistants aux antibiotiques et les entéro-invasifs afin d'espérer diminuer leur incidence en santé publique.

Les pathovars de *E. coli* constituent un groupe de bactéries pathogènes responsables d'un nombre d'infections en constante augmentation. Particulièrement, l'émergence des STEC comme agent majeur de colite hémorragique et du syndrome hémolytique et urémique (SHU) chez les enfants de moins de 15 ans, constitue un réel problème de santé publique. Au niveau socio-économique, ce pathogène présente une importance cruciale, étant donné le nombre de décès humains, les frais d'hospitalisation et les pertes de production engendrées. Les efforts doivent être prioritairement axés dans les secteurs en amont du stade de la distribution et de la consommation. Il est donc indispensable de repérer des réservoirs animaux et d'étudier l'écologie des STEC afin d'améliorer les conditions de production de la filière viande.

Les méthodes de détection des pathovars de *E. coli* étant très complexes et couteuses pour les pays en développement, il serait souhaitable en accord avec les organismes internationaux de trouver un schéma générale de détection et de surveillance de ces pathogènes dans ces pays afin de réduire les peines causées par ces derniers.

De façon générale pour les salmonelles et pathovars de *E. coli*, il faudra promouvoir des démarches intégrées d'analyse de risque reposant sur un partenariat entre les secteurs privés et publics ainsi que les organisations de consommateur. Les contrôles devront être renforcés pour permettre le recueil de données épidémiologiques essentielles, en vue de mieux axer la prévention et l'information.

Ces études ont soulignés l'importance de disposer de données de contamination par les microorganismes pathogènes des aliments. Ces travaux seront poursuivis par des études d'envergures nationales pour mieux comprendre l'épidémiologie des entéropathogènes. Une étude sera réalisée sur toute la chaîne de production des viandes (c'est-à-dire de la ferme à la fourchette) afin de mieux situer les points critiques. Tous ces travaux fourniront des informations scientifiques qui vont attirer l'attention des décideurs politiques et contribuer à des changements dans les pratiques socio-professionnelles des acteurs de la chaîne de production de viandes. Ce qui permettra de lutter efficacement contre les maladies d'origine alimentaire.

REFERENCES

BIBLIOGRAPHIQUES

REFERENCES BIBLIOGRAPHIQUES

Acha PN, Szyfres B. 2001. Zoonoses and Communicable Diseases Common to Man and Animals. Third Edition, Washington DC: *Pan American Health Organization*. 233-246.

Adetosoye AI. 1980. Infectious drug resistance in *E. coli* isolated from livestock. *Zentralbl. Bakteriol. A.* **247**: 25-34.

Akanbi BO, Mbah IP, Kerry PC. 2011. Prevalence of *Escherichia coli* O157:H7 on hides and faeces of ruminants at slaughter in two major abattoirs in Nigeria. *Lett. Appl. Microbiol.* **53**: 336–240.

Aktan I, Sprigings KA, La Ragione RM, Faulkner LM, Paiba GA, Woodward MJ. 2004. Characterization of attaching-effacing *Escherichia coli* isolated from animals at slaughter in England and Wales. *Vet. Microbiol.* **102**: 43–53.

Alonso MZ, Padola NL, Parma AE, Lucchesi PMA. 2011. Enteropathogenic *Escherichia coli* contamination at different stages of the chicken slaughtering process. *Poult. Sci.* **90**: 2638–2641.

Altekruse SF, Street DA, Fein SB, Levy AS. 1996. Consumer knowledge of foodborne microbial hazards and food-handling practices. *J. Food Prot.* **59**: 287–294.

Andrade JR, Da Veiga VF, De Santa Rosa MR, Suassuna I. 1989. An endocytic process in HEp-2 cells induced by enteropathogenic *Escherichia coli*. *J. Med. Microbiol.* **28**: 49–57.

Angelillo IF, Foresta MR, Scozzafava C, Pavia M. 2001. Consumers and foodborne diseases: knowledge, attitudes and reported behavior in one region of Italy. *Int J. Food Microbiol.* **64**: 161–166.

Antikainen J, Tarkka E, Haukka K, Siitonen A, Vaara M, Kirveskari J. 2009. New 16-plex PCR method for rapid detection of diarrheagenic *Escherichia coli* directly from stool samples. *Eur. J. Clin. Microbiol. Infect. Dis.* **28**: 899–908.

APHA (American Public Health Association). 2004. Control of Communicable Diseases Manual, (ed.) Heymann DL, 18th Edition, Washington, DC. 700 p.

Arsenault J, Letellier A, Quessy S, Boulianne M. 2007. Prevalence and risk factors for *Salmonella* and *Campylobacter* spp. carcass contamination in broilers chickens slaughtered in Quebec, Canada. *J. Food Prot.* **70**: 1820–1828.

Baggesen DL, Aarestrup FM. 1998. Characterisation of recently emerged multiple antibiotic-resistant *Salmonella enterica* serovar Typhimurium DT 104 and other multiresistant phage types from Danish pig herds. *Vet. Rec.* **143**: 95-97.

Baldini MM, Kaper JB, Levine MM, Candy DC, Moon HW. 1983. Plasmid-mediated adhesion in enteropathogenic *Escherichia coli. J. Pediatr. Gastroenterol. Nutr.* **2**: 534–538.

Barro N, Bello AR, Savadogo A, Ouattara CAT, Ilboudo AJ, Traoré AS. 2006. Hygienic status assessment of dishwaters, utensils, hands and pieces of money in street foods vending sites in Ouagadougou; Burkina Faso. *Afr. J. Biotechnol.* **5**: 1107–1112.

Barro N, Gamene AA, Itsiembou Y, Savadogo A, Nikiema AP, Ouattara CAT, De Souza CA, Traoré AS. 2007. Street–vended foods improvement: Contamination mechanisms and application of food safety objective strategy: Critical review. *Pakistan J. Nutr.* **6**: 1–10.

Barro N, Nikiéma P, Ouattara CAT, Traoré AS. 2002. Evaluation de l'hygiène et de la qualité microbiologique de quelques aliments de rue et les caractéristiques des consommateurs dans les villes de Ouagadougou et de Bobo-Dioulasso (Burkina Faso). *Rev Sci Tech Sci Santé.* **25**: 7–21.

Barro N, Sangaré L, Tahita M, Ouattara CAT, Traoré AS. 2005. Les principaux agents du péril fécal identifiés dans les aliments de rue et ceux des cantines Burkina Faso et ailleurs et les risques de maladies associées. Colloque Régional scientifique et Pédagogique: Maîtrise de procédés en vue d'améliorer la qualité et la sécurité des aliments, Utilisation des OGM, analyses des risques en Agroalimentaires UO/AUF/GP3A/CIDEFA. Du 8 au 10 novembre 2005 à Ouagadougou, Burkina Faso.

Barro N, Tahita MC, Traore O, Sangare L, De Souza CA, Traore AS. 2008. Risks associated with practices, processes and environment of ready-to-eat and street-vended foods that lead to contamination by common foodborne viruses. *In*: Hygiene and Its Role in Health. Ed. P. L. Aderson and J.P. Lachan: pp: 129-153.

Barro N, Traore AS. 2001. Caractéristiques des vendeurs et des consommateurs, inventaires, salubrité des aliments de rue et état de santé des consommateurs. *Rapport d'études sur les aliments de rue. CRSBAN/SADOC, Université de Ouagadougou*, 49 p.

Bell C, Kyriakides A. 2002. Salmonella *In*: Foodborne Pathogens. Hasards, risk analysis and control.Woodhead Publishing Limited. pp :307–334.

Bello AR. 2004. Evaluation des bonnes pratiques d'hygiène et de fabrication et évolution de la flore microbienne au cours de procédés de préparation et de vente des aliments de rue. *Mémoire de DESS, Université de Ouagadougou*; 75 p.

Bettelheim KA, Beutin L, Gleier K, Pearce JL, Luke RKJ, Zimmermann S. 2003. Serotypes of *Escherichia coli* isolated from healthy infants in Berlin, Germany and Melbourne, Australia. *C.I.M.I.D.* **26**: 55–63.

Beutin L, Marchés O, Bettelheim KA, Gleier K, Zimmermann S, Schmidt H, Oswald E. 2003. HEp-2 cell adherence, actin aggregation, and intimin types of attaching and effacing *Escherichia coli* strains isolated from healthy infants in Germany and Australia. *Infect. Immunity.* **71**: 3995–4002.

Blanco J, Blanco M, Blanco JE, Mora A, Gonzales E, Bernardez MI, Alonzo MP, Coira A, Rodrigues A, Rey J, Alonzo JM, Usera MA. 2003. Verotoxin-producing *Escherichia coli* in Spain: prevalence, serotypes and virulence genes of O157:H7 and non-O157 VTEC in ruminants, raw beef products and humans. *Soc. Exp. Biol. Med.* **228**: 345–351.

Blanco M, Padola NL, Krüger A, Sanz ME, Blanco JE, González EA, Dahbi G, Mora A, Bernárdez MI, Etcheverría AI, Arroyo GH, Lucchesi PM, Parma AE, Blanco J. 2004. Virulence genes and intimin types of Shiga-toxin-producing *Escherichia coli* isolated from cattle and beef products in Argentina. *Int Microbiol.* **7**: 269–276.

Boerlin P, McEwen SA, Boerlin-Petzold F, Wilson JB, Johnson RP, Gyles CL. 1999. Association between virulence factors of Shiga toxin-producing *Escherichia coli* and disease in humans. *J. Clin. Microbiol.* **37**: 497–503.

Bohez L, Ducatelle R, Pasmans F, Botteldoorn N, Haesebrouck F, Van Immerseel F. 2006. *Salmonella enterica* serovar Enteritidis colonization of the chicken caecum requires the *HilA* regulatory protein. *Vet. Microbiol.* **116**: 202–210.

Bonardi S, Brindani F, Pizzin G, Lucidi L, D'Incau M, Liebana E, Morabito S. 2003. Detection of *Salmonella* spp., *Yersinia enterocolitica* and verocytotoxin-producing *Escherichia coli* O157 in pigs at slaughter in Italy. *Int. J. Food. Microbiol.* **15**:101–110.

Bonkoungou IJO, Lienemann T, Martikainen O, Dembelé R, Sanou I, Traoré AS, Siitonen A, Barro N, Haukka K. 2011. Detection of diarrhoeagenic *Escherichia coli* by 16-plex PCR from young children in urban and rural Burkina Faso. *Clin Microbiol Infect*; doi: 10.1111/j.1469-0691.2011.03675.x.

Bornet G. 2000. Le poulet sans salmonelles : Mythe ou réalité? *Rev.med.vet.* **12**: 1083–1094.

Bouvet P. 1995. Salmonelles et Salmonelloses en France. Dans: Sécurité alimentaire du Consommateur (Collection STAA). Moll M. et Moll. Ed. Lavoisier: 1–20.

Boyd EF, Hartl D L. 1998. Chromosomal Regions Specific to Pathogenic Isolates of *Escherichia coli* Have a Phylogenetically Clustered Distribution. *J. Bacteriol.* **180**: 1159–1165.

Brandal LT, Lindstedt BA, Aas L, Stavnes TL, Lassen J, Kapperud G. 2007. Octaplex PCR and fluorescence-based capillary electrophoresis for identification of human diarrheagenic *Escherichia coli* and *Shigella* spp. *J. Microbiol. Methods* **68**: 331–341.

Brenner D, Fanning G, Miklos G, Steigerwalt A. 1973. Polynucleotide sequence relatedness among *Shigella* species. *Int. J. Syst. Bacteriol.* **23**: 1–7.

Brugere H, Bailly JD. 2006. Maîtrise de l'hygiène dans la filière viande d'animaux de boucherie. *Bull. Soc. Fr. Microbiol.* **21**: 110–119.

Bryan FL. 1988. Risk associated with practices, procedures and processes that lead to outbreaks of foodborne diseases. *J. Food Prot.* **51**: 663–673.

Butzler JP. 2004. Campylobacter, from obscurity to celebrity. *Clin. Microbiol. Infect.***10**: 868–876.

Bywater R, Deluyker H, Derooker E, de Jong A, Marion H, McConville V, Rowan T, Shryock T, Shuster D, Thomas V, Valle M, Walters J. 2004. European survey of antimicrobial susceptibility among zoonotic and commensal bacteria isolated from food-producing animals. *J. Antimicrob.Chemother.* **54**: 744-754.

Caprioli A, Morabito S, Brugère H, Oswald E. 2005. Enterohaemorrhagic *Escherichia coli*: emerging issues on virulence and modes of transmission. *Vet. Res.* **36**: 289–311.

Cardinale E, Perrier Gros-Claude JD, Tall F, Gueye EF, Salvat G. 2005. Risk factors for contamination of ready-to-eat street-vended poultry dishes in Dakar, Senegal. *Int. J. Food Microbiol.* **103**: 157–165.

Cardinale E, Tall F, Guèye EF, Cissé M, Salvat G. 2004. Risk factors for *Salmonella enterica* subsp. *enterica* infection in Senegalese broiler-chicken flocks. *Prev. Vet. Med.* **63**: 151–161.

Carlier V, Lagrange P. 2001. *Salmonella*, service d'information alimentaire, H.C.S. International. Paris. pp: 84.

Cassels FJ, Wolf MK. 1995. Colonization factors of diarrheagenic *E. coli* and their intestinal receptors. *J. Indian Microbiol.* **15**: 214–226.

Catry B, Laevens H, Devriese LA, Opsomer G, De Kruif A. 2003. Antimicrobial resistance in livestock. *J. Vet. Pharmacol. Ther.* **26**: 81–93.

Catsaras M. 1973. Les intoxications alimentaires par les viandes et produits carnés. *Ind. Aliment. Agric.* **90**: 1467-1473.

CDC. 1994. African pygmy hedgehog-associated salmonellosis,Washington, 1994. *MMWR Morb. Mortal. Wkly Rep.* **44**: 462–463.

CDC. 2001. Diagnosis and management of foodborne illnesses: A primer for physicians. *MMWR Recomm. Rep.* **50**: 1–69.

CDC. 2008. *Salmonella* surveillance: annual summary, 2006. Centers for Disease Control and Prevention, Atlanta, GA. **http://www.cdc.gov/ncidod /dbmd/phlisdata/salmonella.htm.**

Chahed A. 2008. Prévalence et caractérisation de souches d'*Escherichia coli* O157 producteurs de shiga-toxines isolées de denrées alimentaires d'origine animale en Belgique et en Algérie. *Med. Vet.***152**: 39–43.

Chakravarty I., Canet C. 1996. Street-food in Calcutta. *Food, Nut. Agri.* **17/18**: 7p

Chambers JR, Bisaillon JR, Labbe Y, Poppe C, Langford CF. 1998. *Salmonella* prevalence in crops of Ontario and Quebec broiler chickens at slaughter. *Poult. Sci.* **77**: 1497–1501.

Cheftel E, Spiegel A, Bornet G, Morell E, Michel E, Buisson Y. 1997. Toxic food infection caused by Shigella flexneri in a military unit. *Cah. Santé,* **7**: 295–299.

Chiu CH, Chen HL, Kao LS, Yang CY, Chu C, Doublet B, Praud K, Cloeckaert A. 2007. Variant *Salmonella* genomic island 1 antibiotic resistance gene clusters in *Salmonella* enterica serovar Derby isolates from humans in Taiwan. *J. Antimicrob. Chemother.* **59**: 325–326.

Clark, GM, Kaufmann AF, Gangarosa EJ, Thompson MA. 1973. Epidemiology of an international outbreak of *Salmonella* Agona. *Lancet.* **2**: 490–493.

Clermont O, Olier M, Hoede C, Diancourt L, Brisse S, Keroudean M, Glodt J, Picard B, Oswald E, Denamur E. 2011. Animal and human pathogenic *Escherichia coli* strains share common genetic backgrounds. *Infect. Genet. Evol.* **11**: 654–662.

Clinical and Laboratory Standards Institute (CLSI). 2009. Methods for Dilution Antimicrobial Susceptibility Tests for Bacteria that Grow Aerobically. Available online at: **http://www.clsi.org/source/orders/free/m07-a8.pdf. Accessed 1. December 2011**.

Codex Alimentarius **Commission.** 2005. Code of hygienic practice for meat. Adopted by the CAC, New Zealand, July 2005 (CAC/RCP 58-2005), 52p.

Cooke EM. 1985. *Escherichia coli*--an overview. *J. Hyg.* **95**: 523–530.

Cookson AL, Bennett J, Thomson-Carter F, Attwood GT. 2007. Molecular subtyping and genetic analysis of the enterohemolysin gene (*ehxA*) from Shiga toxin-producing *Escherichia coli* and atypical enteropathogenic *E. coli*. *Appl. Environ. Microbiol.* **73**: 6360–6369.

Cookson AL, Cao M, Bennett J, Nicol C, Thomson-Carter F, Attwood GT. 2010. Relationship between virulence gene profiles of atypical enteropathogenic *Escherichia coli* and Shiga toxin-producing *E. coli* isolates from cattle and sheep in New Zealand. *Appl. Environ. Microbiol.* **76**: 3744–3747.

Cookson AL, Taylor SCS, Bennett J, Thomson-Carter F, Attwood GT. 2006. Serotypes and analysis of distribution of Shiga toxin-producing *Escherichia coli* from cattle and sheep in the lower North Island, New Zealand. *New Zealand Vet. J.* **54**:78–84.

Cookson ST, Nataro JP. 1996. Characterization of HEp-2 cell projection formation induced by diffusely adherent *Escherichia coli*. *Microb. Pathol.* **21**: 421–434.

Courvalin P, Phillippon A. 1989. Mécanismes biochimiques de la résistance bactérienne aux agents antibactériens. *In* : Le Minor; Véron M. *Bact. Méd.* **14**: 332–355.

Das SC, Khan A, Panja P, Datta S, Sikdar A, Yamasaki S, Takeda Y, Bhattacharya SK, Ramamurthy T, Balakrish nair G. 2005. Dairy farm investigation on Shiga toxin-producing *Escherichia coli* (STEC) in Kolkata, India with emphasis on molecular characterization. *Epidemiol. Infect.* **133**: 617–626.

Davies PR, Bovee FG, Funk JA, Morrow WE, Jones FT, Deen J. 1998. Isolation of *Salmonella* serotypes from feces of pigs raised in a multiple-site production system. *J. Am. Vet. Med. Assoc.* **212**: 1925–1929.

Dibner JJ, Richards JD. 2005. Antibiotic growth promoters in agriculture: history and mode of action. *Poult. Sci.* **84**: 634–643.

Dione MM, Ieven M, Garin B, Marcotty T, Geerts S. 2009. Prevalence and antimicrobial resistance of *Salmonella* isolated from broiler farms, chicken carcasses, and street-vended restaurants in Casamance, Senegal. *J. Food Prot.* **72**: 2423–2427.

Dione MM, Ikumapayi UN, Saha D, Mohammed NI, Geerts S, Ieven M, Adegbola RA, Antonio M. 2011. Clonal Differences between Non-Typhoidal *Salmonella* (NTS) Recovered from Children and Animals Living in Close Contact in The Gambia. *PLoS Negl. Trop. Dis.* **5**: e1148. doi:10.1371/journal.pntd.0001148.

Dipineto L, Santaniello A, Fontanella M, Lagos K, Fioretti A, Menna LF. 2006. Presence of Shigatoxin-producing *Escherichia coli* O157:H7 in living layer hens. *Lett. Appl. Microbiol.* **43**: 293–295.

Drabo B, Poda A, Yaméogo F. 2001. Gestion de la faune et des ressources agropastorales: Cas de la zone de Darkoye, Burkina Faso Contribution pour atelier " faune sauvage et bétail: complémentarité, coexistence ou compétition ? " Niamey 16-19 Janvier 2001. 2 p.

Edlund C, Nord CE. 2000. Effect on the human normal microflora of oral antibiotics for treatment of urinary tract infections. *J. Antimicrob. Chemother.* **1**: 41–48.

Edrington TS, Hume ME, Looper ML, Schultz CL, Fitzgerald AC, Callaway TR, Genovese KJ, Bischoff KM, McReynolds JL, Anderson RC, Nisbet DJ. 2004. Variation in the faecal shedding of *Salmonella* and *E. coli* O157:H7 in lactating dairy cattle and examination of *Salmonella* genotypes using pulsed-field gel electrophoresis. *Lett. Appl. Microbiol.* **38**: 366–372.

Eisenstein BI. 1996. Fimbriae. In: *Escherichia coli* and *Salmonella* Typhimurium, Neidhart FC.eds. Washington D.C. American Society for Microbiology. pp: 84–90.

Elder RO, Keen JE, Siragusa GR, Barkocy-Gallagher GA, Koohmaraie M, Laegreid WW. 2000. Correlation of enterohemorrhagic *Escherichia coli* O157 prevalence in feces, hides, and carcasses of beef cattle during processing. *Proc. Nat. Acad. Sci. USA*. **97**: 2999–3003.

Escobar-Paramo P, Clermont O, Blanc-Potard A-B, Bui H, Le Bouguenec C, Denamur E. 2004. A Specific Genetic Background Is Required for Acquisition and Expression of Virulence Factors in *Escherichia coli*. *Mol. Biol. Evol.* **21**: 1085-1094.

Estrada-Garcia T, Lopez-Saucedo C, Zamarripa-Ayala B, Thompson MR, Gutierez-Cogo L, Perez-Martinez I, Escobar-Gutierrez A. 2004. Prevalence of *Escherichia coli* and *Salmonella* spp. In street-vended food of open markets (tianguis) and general hygienic and trading practices in Mexico City. *Epidemiol. Infect.* **132**: 1181–1184.

Ethelberg S, Smith B, Torpdahl M, Lisby M, Boel J, JensenT, Molbak K. 2007. An outbreak of Verocytotoxin-producing *Escherichia coli* O26:H11 caused by beef sausage, Denmark 2007. *Euro Surveill.* **12**: 70531–70534.

European Food Safety Authority (EFSA). 2011. Urgent advice on the public health risk of Shiga-toxin producing *Escherichia coli* in fresh vegetables. *EFSA J.* **9**: 2274.

Euzeby JP. 1996. Les Salmonelles et Salmonelloses dûes aux sérovars ubiquistes. Compte rendu d'intervention lors du séminaire de l'UCAAB, nov.1996.

Fagan PK, Hornitzky MA, Bettelheim KA, Steven PD. 1999. Detection of Shiga-Like Toxin (stx1 and stx2), Intimin (eaeA), and Enterohemorrhagic *Escherichia coli* (EHEC) Hemolysin (EHEC hlyA) Genes in Animal Feces. *Appl. Environ. Microbiol.* **65**: 868–872.

FAO. 2005. Code of Hygienic Practice for Meat CAC/RCP 58-200, 2005.

FAO/OMS. 2004. Programme mixte FAO/OMS sur les normes alimentaires commission du Codex Alimentarius. Vingt-septième session Genève (Suisse), 28 juin - 3 juillet 2004. Rapport de la trente sixième session du comité du codex sur les additifs alimentaires et les contaminants Rotterdam (Pays-Bas) 22 - 26 mars 2004. 223p.

Farooq S, Hussain I, Mir MA, Bhat MA, Wani SA. 2009. Isolation of atypical enteropathogenic *Escherichia coli* and Shiga toxin 1 and 2f-producing

Escherichia coli from avian species in India. *Lett. Appl. Microbiol.* **48**: 692–697.

Fashae K, Ogunsola F, Aarestrup F M, Hendriksen RS. 2010. Antimicrobial susceptibility and serovars of *Salmonella* from chickens and humans in Ibadan, Nigeria. *J. Infect. Dev. Ctries.* **4**: 484–494.

Fone DL, Barker RM. 1994. Associations between human and farm animal infections with *Salmonella typhimurium* DT104 in Herefordshire. *Commun. Dis. Rep. CDR Rev.* **4**: R136-R140.

François J, Monique C, Michèle W, Alain G. 2003. De l'antibiogramme à la prescription. Edition Biomérieux 2e édition, 29p.

Fratamico PM, Bhagwat AA, Injaian L, Fedorka-Cray PJ. 2008. Characterization of Shiga toxin-producing *Escherichia coli* strains isolated from swine feces. *Foodborne Pathog. Dis.* **5**: 827–838.

Garcia-Fernandez A, Fortini D, Veldman K, Mevius D, Carattoli A. 2008. Characterization of plasmids harbouring qnrS1, qnrB2 and qnrB19 genes in *Salmonella*. *J Antimicrob Chemother* **63**: 274-281.

Gaul SB, Wedel S, Erdman MM, Harris DL, Harris IT, Ferris KE, Hoffman L. 2007. Use of pulsed-field gel electrophoresis of conserved XbaI fragments for identification of swine *Salmonella* serotypes. *J. Clin. Microbiol.* **45**: 472–476.

Ghafi Y, Daube G. (2007). Le point sur les méthodes de surveillance de la contamination microbienne des denrées alimentaires d'origine animale. *Ann. Méd. Vet.* **151**: 79–100.

Gillespie IA, O'Brien SJ, Adak GK, Ward LR, Smith HR. 2005. Foodborne general outbreaks of *Salmonella* Enteritidis phage type 4 infection, England and Wales, 1992–2002: where are the risks? *Epidemiol. Infect.* **133**: 795–801.

Gillespie IA, O'Brien SJ, Adak GK, Ward LR, Smith HR. 2005. Foodborne general outbreaks of *Salmonella* Enteritidis phage type 4 infection, England and Wales, 1992–2002: where are the risks? *Epidemiol Infect.* **133**:795–801.

Girardeau JP, Dalmasso A, Bertin Y, Ducrot C, Bord S, Livrelli V, Vernozy-Rozand C, Martin C. 2005. Association of Virulence Genotype with

Phylogenetic Background in Comparison to Different Seropathotypes of Shiga Toxin-Producing *Escherichia coli* Isolates. *J. Clin. Microbiol.* **43**: 6098–6107.

Glaser CA, Angulo FJ, Rooney J. 1994. Animal associated opportunistic infections in HIV-infected persons. *Clin. Infect. Dis.* **18**: 14–24.

Greatorex JS, Thorne GM. 1994. Humoral immune responses to Shiga-like toxins and *Escherichia coli* O157 lipopolysaccharidein hemolytic-uremic syndrome patients and healthy subjects. *J. Clin. Microbiol.* **32**: 1172–1178.

Green SDR, Cheesbrough JS. 1993. *Salmonella* bacteraemia among young children at a rural hospital in western Zaire. *Ann. Trop. Paediatr.* **13**: 45-53.

Griffin PM, Tauxe RV. 1991. The epidemiology of infections caused by *Escherichia coli* O157:H7, other enterohemorrhagic *E. coli*, and the associated hemolytic uremic syndrome. *Epidemiol. Rev.* **13**: 60–97.

Griffin PM. 1995. *Escherichia coli* O157:H7 and other enterohemorrhagic *Escherichia coli.. In*: Blaser, M.J., Smith, P.D., Ravdin, J.I., Greenberg, H.B., Guerrant, R.L., (Eds.), Infections of the gastrointestinal tract. Raven Press, Ltd., New York, N.Y. pp: 739–761

Grimont P. 1987. Taxonomie des *Escherichia*. *Méd. Mal. Infect.* (**Numéro spécial**): 6-1.

Grimont PAD, Grimont F, Bouvet P. 2000. Molecular basis of the diversity in the genus *Salmonella*. *In*: *Salmonella* in domestic animals.Wray et al. CABI Publishing, British Library, London, U.K. pp 1–17.

Grimont PAD. 1992. Les marqueurs épidémiologiques des Salmonella. *Méd. Mal. Inf.* **22**: 249–257.

Gyles CL. 2007. Shiga toxin-producing *Escherichia coli*: an overview. *J. Anim. Sci.* **85**: 45–62.

Hald T, Lo Fo Wong DM, Aarestrup FM. 2007. The attribution of human infections with antimicrobial resistant *Salmonella* bacteria in Denmark to sources of animal origin. *Foodborne Pathog. Dis.* **4**: 313–326.

Hald T, Vose D, Wegener HC, Koupeev T. 2004. A Bayesian approach to quantify the contribution of animals-food sources to human salmonellosis. *Risk Anal.* **24**: 255–269.

Harbottle H, White DG, McDermott PF, Walker RD, Zhao S. 2006. Comparison of multilocus sequence typing, pulsed-field gel electrophoresis, and antimicrobial susceptibility typing for characterization of *Salmonella enterica* serotype Newport isolates. *J Clin Microbiol.* **44**:2449–2457.

Herrero A, Rodicio, MR, Gonzalez-Hevia, MA, Mendoza MC. 2006. Molecular epidemiology of emergent multidrug-resistant *Salmonella enterica* serotype Typhimurium strains carrying the virulence resistance plasmid pUO-StVR2. *J. Antimicrob. Chemother.* **57**: 39–45.

Heuvelink AE, Zwartkruis-Nahuis JT, van den Biggelaar FL, van Leeuwen WJ, de Boer E. 1999. Isolation and characterization of verocytotoxin-producing *Escherichia coli* O157 from slaughter pigs and poultry. *Int. J. Food Microbiol.* **52**: 67–75.

Heyderman RS, Soriani M, Hirst TR. 2001. Is immune cell activation the missing link in the pathogenesis of post-diarrhoeal HUS? *Trends Microbiol.* **9**: 262-266.

Heyndrickx M, Pasmans F, Ducatelleb R, Decostere A, Haessebrouck. 2005. Recent changes in *Salmonella* nomenclature: The need for clarification. *Vet. J.* **170**: 275–277.

Hu L, Kopecko DJ. 2003. *Campylobacter* Species. *In*: Miliotis MD, Bier JW. (Ed.) International Handbook of Foodborne Pathogens. Marcel Dekker: New York. pp 181–198.

Huang DB, Okhuysen PC, Jiang Z, Dupont HL. 2004. Enteroaggregative *Escherichia coli*: an emerging enteric pathogen. *Am. J. Gastroenterol.* **99**: 383–389.

Humbert, F. 1998. Les salmonelles. *In*: Manuel de bactériologie alimentaire. Polytechnica (ed): 27-52.

Humphrey T. 2000. Public-health aspects of *Salmonella* infection. In: *Salmonella* in Domestic Animals. Way C and Way A (eds.). Oxon, United Kingdom: CABI Publishing. pp 245–263.

Hussein HS, Sakuma T. 2005. Prevalence of Shiga toxin producing *Escherichia coli* in dairy cattle and their products. *J. Dairy Sci.* **88**:450-465.

Hussein HS. 2007. Prevalence and pathogenicity of shiga toxin-producing *Escherichia coli* in beef cattle and their products. *J. Anim. Sci.* **85**: 63–72.

Ishihara K, Takahashi T, Morioka A, Kojima A, Kijima, Asai T, Tamura Y. National surveillance of *Salmonella enterica* in food-producing animals in Japan. *Acta Veterinaria Scandinavica.* **51**:35 doi:10.1186/1751-0147-51-35

Islam MA, Mondol AS, Azmi IJ, de Boer E, Beumer RR, Zwietering MH, Heuvelink AE, Talukder KA. 2010. Occurrence and characterization of Shiga toxin-producing *Escherichia coli* in raw meat, raw milk, and street vended juices in Bangladesh. *Foodborne Pathog. Dis.* **7**: 1381–1385.

Islam MA, Mondol AS, de Boer E, Beumer RR, Zwietering MH, Talukder KA, Heuvelink AE. 2008. Prevalence and Genetic Characterization of Shiga Toxin-Producing *Escherichia coli* Isolates from Slaughtered Animals in Bangladesh. *Appl. Environ. Microbiol.* **74**: 5414–5421.

ISND. 2006. Rapport sur le recensement general de la population et de l'habitation. Ministère de l'économie et des finances, Burkina Faso. 52p.

Jang HY, Lee SJ, Lim GJ, Lee SH, Kim JT, Park HJ, Chung HB, Choe HN. 2008. The rate of *Salmonella* spp. infection in zoo animals at Seoul Grand Park, Korea. *J. Vet. Sci.* **9**: 177–181.

Jenkins C, Gillespie SH. 2006. Principles and Practice of Clinical Bacteriology *Salmonella* spp. 2nd ed. Editors Stephen H, Gillespie, Peter M. Hawkey © John Wiley & Sons, Ltd. Pp 367–376.

Jerse AE, Yu J, Tall BD, Kaper JB. 1990. A genetic locus of enteropathogenic *Escherichia coli* necessary for the production of attaching and effacing lesions on tissue culture cells. *Proc Natl. Acad. Sci. U S A.* **87**: 7839–7843.

Joffin C, Joffin N. 1994. Microbiologie alimentaire 4ème éd : Centre Régional de Documentation Pédagogique, 204p.

Johnsen G, Wasteson Y, Heir E, Berget OI, Herikstad H. 2001. *Escherichia coli* O157:H7 in faeces from cattle, sheep and pigs in the southwest part of Norway during 1998 and 1999. *Int. J. Food Microbiol.* **65**: 193–200.

Jorgensen F, Bailey R, Williams S, Hendersonp, Wareing DR, Bolton FJ, Frost JA, Ward L, Humphrey TJ. 2002. Prevalence and numbers of *Salmonella* and *Campylobacter* spp. on raw, whole chickens in relation to sampling methods. *Int. J. Food Microbiol.* **76**: 151–164.

Kagambèga A, Barro N, Traoré AS, Siitonen A, Haukka K. 2012. Characterization of *Salmonella enterica* and diarrheagenic *Escherichia coli* from poultry carcasses sold at open markets in Ouagadougou, Burkina Faso. *Foobborne Pathog. Dis.*

Kagambega A, Haukka K, Siitonen A, Traoré AS, Barro N. 2011. Prevalence of *Salmonella enterica* and the hygienic indicator *Escherichia coli* in raw meat at markets in Ouagadougou, Burkina Faso. *J. Food Prot.* **74**: 1547–1551.

Kagambèga A, Martikainen O, Lienemann T, Siitonen A, Traoré A, Barro N, Haukka K. 2012. Diarrheagenic *Escherichia coli* detected by 16-plex PCR in raw meat and beef intestines sold at local markets in Ouagadougou, Burkina Faso. *Int. J. Food Microbiol.* **153**: 154–158.

Kaper JB, Nataro JP, Mobley HL. 2004. Pathogenic *Escherichia coli*. *Nature Reviews Microbiol.* **2**: 123–1240.

Kariuki S, Gilks C, Kimari J, Muyodi J, Waiyaki P, Hart CA. 1999. Analysis of *Salmonella enterica* serotype Typhimurium by phage typing, antimicrobial susceptibility and pulsed-field gel electrophoresis. *J Med Microbiol.* **48**: 1037–1042.

Kariuki S, Gilks C, Kimari J, Obanda A, Muyodi J, Waiyaki P, Hart CA. 1999. Genotype analysis of *Escherichia coli* strains isolated from children and chickens living in close contact. *Appl. Environ. Microbiol.* **65**: 472–476.

Kariuki S. 2008. Typhoid fever in sub-Saharan Africa: Challenges of diagnosis and management of infections. *J. Infect. Developing Countries.* **2**: 443–447.

Karmali MA, Gannon V, Sargeant JM. 2010. Verocytotoxin-producing *Escherichia coli* (VTEC). *Vet. Microbiol.* **140**: 360–370.

Kauffmann F. 1971. Classification and nomenclature of the genus *Salmonella. Acta Pathol Microbiol Scand [B] Microbiol. Immunol.* **79**: 421–422.

Kegode RB, Doetkott DK, Khaitsa ML, Wesley IV. 2008. Occurrence of *Campylobacter* species, *Salmonella* species and generic *Escherichia coli* in meat products from retail outlets in the Fargo metropolitan area. *J. Food Safety.* **28**: 111–125.

Keskimäki M, Mattila L, Peltola H, Siitonen A. 2000. Prevalence of diarrheagenic *Escherichia coli* in Finns with or without diarrhea during a round-the-world trip. *J. Clin. Microbiol.* **38**: 4425–4429.

Kikuvi GM, Ombui JN, Mitema ES. 2010. Serotypes and antimicrobial resistance profiles of *Salmonella* isolates from pigs at slaughter in Kenya. *J Infect Dev Ctries.* **4:** 243-248.

King LK, Awumbila B, Canacoo EA, Ofosu-Amaah S. 2000. An assessment of the safety of street foods in the Ga district of Ghana; implication for the spread of zoonoses. *Acta Tropica.* **76**: 39–43.

Kobayashi H, Pohjanvirta T, Pelkonen S. 2002. Prevalence and characteristics of intimin and Shiga toxin-producing *Escherichia coli* from gulls, pigeons and broilers in Finland. *J. Vet. Med. Sci.* **64**: 1071–1073.

Kobayashi H, Shimada J, Nakazawa M, Morozumi T, Pohjanvirta T, Pelkonen S, Yamamoto K. 2001. Prevalence and characteristics of Shiga toxin-producing *Escherichia coli* from healthy cattle in Japan. *Appl. Environ. Microbiol.* **67**: 484–489.

Krause G, Zimmermann S, Beutin L. 2005. Investigation of domestic animals and pets as a reservoir for intimin- (eae) gene positive *Escherichia coli* types. *Vet. Microbiol.* **106**: 87–95.

Krauss H, Weber A, Appel M, Enders B, Isenberg HD, Schiefer HG, Slenczka W, Von Graevenitz A, Zahner H. 2003. Zoonoses: infectious diseases transmissible from animals to humans. ASM Press: Washington. 456 p.

Lagrance P, Reinert P. 1987. L'antibiothérapie en pratique quotidienne. Centre de documentation des laboratoires BRISTOL, 64.

Lamont RJ. 2004. Advances in Molecular and Cellular Microbiology 5 Bacterial Invasion of Host Cells EDITED BY Lamont University of Florida© Cambridge University Press.

Lan R, Alles MA, Donohoe K, Martinez MB, Reeves PR. 2004. Molecular evolutionary relationships of enteroinvasive *Escherichia coli* and *Shigella* spp. *Infect Immun.* **72**: 5080–5088.

Le Minor L, Popoff MY, Bockemuhl J. 1990. Supplement 1989 to the Kauffmann-White scheme. *Research Microbiol.* **141**: 1173–1177.

Le Minor L, Popoff MY, Laurent B, Hermant D. 1986. Individualisation d'une septième sous-espèce de *Salmonella*: *S. choleraesuis subsp.indica* subsp. nov. *Ann Inst Pasteur* (Paris). **137**B: 211–217.

Le Minor L, Popoff MY. 1987. Request for an opinion. Designation of *Salmonella enterica* sp. nov., nom. rev., as the type and only species of the genus *Salmonella*. *Int. J. Syst. Bacterial.* **37**: 465–468.

Le Minor L, Rohde R, Taylor J. 1970. Nomenclature of *Salmonella*. *Ann Inst Pasteur* (Paris). **119**: 206–210.

Le Minor L, Veron M, Popoff MY. 1982. Taxonomie des *Salmonella*. *Ann Microbiol.* **133**B: 223–243.

Le Minor L. 1994. The genus *Salmonella*. In: The procaryotes.Ballows and all.Springer, New York: pp 2760–2774.

Lee GY, Jang HI, Hwang IG, Rhee MS. 2009. Prevalence and classification of pathogenic *Escherichia coli* isolated from fresh beef, poultry, and pork in Korea. *Int. J. Food Microbiol.* **134**: 196–200.

Leegaard TM, Van Gestel MH, Petit PLC, Van de Klundert JAM. 1996. Antibiotic resistance mechanisms in *Salmonella* species causing bacteraemia in Malawi and Kenya. *APMZS*. **104**: 302-306.

Lepage P, Bogaerts J, Nsengumuremyi F, Hitimana DG, Van Goethem C, Vandepitte J, Butzler JP. 1984. Severe multiresistant *Salmonella typhimurium* systemic infections in Central Africa - clinical features and treatment in a paediatric department. *J Antimicrob Chemother*. **14**: 153-159.

Levine MM, Edelman R. 1984. Enteropathogenic *Escherichia coli* of classic serotypes associated with infant diarrhea: epidemiology and pathogenesis. *Epidemiol. Rev.* **6**: 31–51.

Levine MM. 1987. *Escherichia coli* that cause diarrhea: enterotoxigenic, enteropathogenic, enteroinvasive, enterohemorrhagic, and enteroadherent. *J. Infect. Dis.* **155**: 377–389.

Leyral G. 1996. Microbiologie et Toxicologie des aliments : hygiène et sécurité alimentaires. Collection "Bioscience et techniques": Centre Régional de Documentation Pédagogique d'Aquitaine. 266p.

Liesegang A, Tschäpe H. 2002. Modified pulsed-field gel electrophoresis method for DNA degradation-sensitive *Salmonella enterica* and *Escherichia coli* strains. *Int J Med Microbiol*. **291**: 645-648.

Lindqvist N, Siitonen A, Pelkonen S. 2002. Molecular Follow-Up of *Salmonella enterica* subsp. *enterica* Serovar Agona Infection in Cattle and Humans. *J. Clin. Microbiol.* **40**: 3648–3653.

Lipsky S, Tanino T. 1995. African pygmy hedgehog-associated salmonellosis Washington, 1994. *Morbid. Mortal. Weekly Rep.* **44**:462–463.

Lopez-Saucedo C, Cerna JF, Villegas-Sepulveda N, Thompson R, Velazquez FR, Torres J, Tarr PI, Estrada-Garcia T. 2003. Single multiplex polymerase reaction to detect diverse loci associated with diarrheagenic *Escherichia coli*. *Emerg. Infect. Dis.* **9**: 127–131.

Low JC, Angus M, Hopkins G, Munro D, Rankin SC. 1997. Antimicrobial resistance of *Salmonella enterica* Typhimurium DT 104 isolates and investigation of strains with transferable apramycin resistance. *Epidemiol Znfect.* **118**: 97-103.

Madoroba E, Van Driessche E, De Greve H, Mast J, Ncube I, Read J, Beeckmans S. 2009. Prevalence of enterotoxigenic *Escherichia coli* virulence genes from scouring piglets in Zimbabwe. *Trop. Anim. Health Prod.* **41**:1539–1547.

Majalija S, Segal H, Ejobi F, Elisha BG. 2008. Shiga toxin gene-containing *Escherichia coli* from cattle and diarrheic children in the pastoral systems of southwestern Uganda. *J. Clin. Microbiol.* **46**: 352–354.

McEvoy JM, Doherty AM, Sheridan JJ, Blair IS, McDowell DA. 2003. The prevalence of *Salmonella* spp. in bovine faecal, rumen and carcass samples at a commercial abattoir. *J. Appl. Microbiol.* **94**: 693–700.

Mead GC. 1993. Problems of producing safe poultry. *Royal Soc. Med. J.* **85**: 39–42.

Mead PS, Slutsker L, Dietz V, McCaig LF, Bresee JS, Shapiro C, Griffin PM and Tauxe RV. 1999. Food-related illness and death in the United States. *Emerg. Infect. Dis.* **5**: 607-625.

Medeiros MI, Neme SN, da Silva P, Capuano DM, Errera MC, Fernandes SA, do Valle GR, de Avila. FA. 2001. Etiology of acute diarrhea among children in Ribeiro Preto-SP, Brazil. *Revista do Instituto de Medicina Tropical de Sao Paulo.* **43**: 21–24.

Mellmann A, Harmsen D, Cummings CA, Zentz EB, Leopold SR, Rico A, Prior K, Szczepanowski R, Ji Y, Zhang W, McLaughlin SF, Henkhaus JK, Leopold B, Bielaszewska M, Prager R, Brzoska PM, Moore RL, Guenther S, Rothberg JM, Karch H. 2011. Prospective Genomic Characterization of the German Enterohemorrhagic *Escherichia coli* O104:H4 Outbreak by Rapid Next Generation Sequencing Technology. PLoS One 6, e22751.

Meng J, Doyle MP, Zhao T, Zhao S. 2007. Enterohemorrhagic *Escherichia coli*, p. 249–269. In. Doyle MP, Beuchat LR. (eds.) Food Microbiology: Fundamentals and Frontiers. 3. ed. American Society for Microbiology. Washington, DC.

Meng J, Zhao S, Doyle MP. 1998. Virulence genes of Shiga-toxin producing *Escherichia coli* isolated from food, animals and humans. *Int J Food Microbiol.* **45**: 229–235.

Mensah P, Armar-Klemesu M, Hammond AS, Haruna A, Nyarko R. 2001. Bacterial contaminants in lettuce, tomatoes, beef and goat meat from metropolitan Accra. *Ghana Med. J.* **35**: 1-6.

Mensah P, Yeboah-Manu D, Owusu-Darko K, Ablordey A. 2002. Street foods in Accra, Ghana: how safe are they? *Bull. World Health Organ.* **80**: 546–554.

Mersha G, Asrat D, Zewde BM, Kyule M. 2010. Occurrence of *Escherichia coli* O157:H7 in faeces, skin and carcasses from sheep and goats in Ethiopia. *Lett. Appl. Microbiol.* **50**: 71–76.

Michael GB, Cardoso M, Schwarz S. 2006. Molecular analysis of *Salmonella enterica* subsp. *enterica* serovar Agona isolated from slaughter pigs. *Vet. Microbiol.* **112**: 43–52.

Millemann Y. 1998. Le pouvoir pathogène des salmonelles: facteurs de virulence et modèles d'étude. *Vet. Res.* **29**: 385–407.

Ministère de la santé. 2007. Tableau de bord santé 2006. Direction des études et de la planification. Ministère de la santé-Burkina Faso, pp. 61.

Mochizuki Y, Masuda H, Kanazashi S, Hosoki Y, Itoh K, Ohishi K, Nishina T, Handa Y, Shiozawa K, Miwa Y. 1992. Clinical and epidemiological aspects of enteritis due to *Salmonella* Hadar: I. Isolation of *S. hadar* from sporadic diarrhea-clinical and bacteriological study. *J. Japanese Assoc. Infect. Dis.* **66**: 22–29.

Molbak K. 2005. Human health consequences of antimicrobial drug-resistant *Salmonella* and other foodborne pathogens. *Clin. Infect. Dis.* **41**: 1613–1620.

Molla B, Alemayehu D, Salah W. 2003. Sources and distribution of *Salmonella* serotypes isolated from food animals, slaughterhouse personnel and retail meat products in Ethiopia: 1997-2002. *Ethip. J. Health Dev.* **17**: 63–70.

Morpeth SC, Ramadhani HO, Crump JA. 2009. Invasive non-Typhi Salmonella disease in Africa. *Clin. Infect. Dis.* **49**: 606–611.

Müller D, Greune L, Heusipp G, Karch H, Fruth A, Tschäpe H, Schmidt HA. 2007. Identification of unconventional intestinal pathogenic *Escherichia coli* isolates expressing intermediate virulence factor profiles by using a novel single-step multiplex PCR. *Appl. Environ. Microbiol.***73**: 3380–3390.

Nataro JP, Kaper JB. 1998. Diarrheagenic *Escherichia coli*. Clin. Microbiol. Rev. 11: 142–201.

NCCLS. 2003. Performance standards for antimicrobial disk susceptibility tests: approved standard M2-A8, 8th ed. National Committee for Clinical Laboratory Standards, Wayne PA.

Nzouankeu A, Ngandjio A, Ejenguele G, Njine T, Wouafo NM. 2010. Multiple contaminations of chickens with *Campylobacter, Escherichia coli* and *Salmonella* in Yaounde (Cameroon). *J. Infect. Developing Countries.* **4**: 583–686.

Obi CL, Bessong PO. 2002. Diarrhoeagenic bacterial pathogens in HIV-positive patients with diarrhoea in rural communities of Limpopo Province, South Africa. *J. Health Popular Nutr.* **20**: 230–234.

Ochman H, Selander RK. 1984. Standard Reference Strains of *Escherichia coli* from Natural Populations. *J. Bacteriol.* **157**: 690–693.

Ojo OE, Ajuwape ATP, Otesile EB, Owoade AA, Oyekunle MA, Adetosoye AI. 2010. Potentially zoonotic shiga toxin-producing *Escherichia coli* serogroups in the faeces and meat of food-producing animals in Ibadan, Nigeria. *Int. J. Food Microbiol.* **142**: 214–221.

Okeke IN. 2009. Diarrheagenic *Escherichia coli* in sub-Saharan Africa: status, uncertainties and necessities. *J. Infect. Dev. Ctries.* **3**: 817–842.

OMS. 2007. Salubrité des aliments et maladies d'origine alimentaire. Aide-mémoire N°237.

Oosterom J. 1991. Epidemiological studies and proposed preventive measures in the fight against human salmonellosis. *Int. J. Food Microbial.* **12**: 41–52.

Ostroff S, Tarr P, Neill M, Lewis J, Hargrett-Bean N, Kobayashi J. 1989. Toxin genotypes and plasmid profiles as determinants of systemic sequelae in *Escherichia coli* O157:H7 infections. *J. Infect. Dis.* **160**: 994–999.

Parker CT, Huynh S, Quinones B, Harris LJ, Mandrell RE. 2010. Comparison of genotypes of *Salmonella enterica* serovar Enteritidis phage type 30 and 9c strains isolated during three outbreaks associated with raw almonds. *Appl. Environ. Microbiol.* **76**: 3723–3731.

Paton JC, Paton AW.1998. Pathogenesis and diagnosis of Shiga toxin-producing *Escherichia coli* infections. *Clin. Microbiol. Rev.* **11**: 450–479.

Pearson AD, Greenwood M, Healing TD, Rollins D, Shahamat M, Donaldson J, Colwell RR, 1993. Colonization of broiler chickens by waterborne *Campylobacter jejuni. Appl. Environ. Microbiol.* **59**: 987–996.

Pedersen K, Lassen-Nielsen AM, Nordentoft S, Hammer AS. 2009. Serovars of *Salmonella* from captive reptiles. *Zoonoses Public Health.* **56**: 238–242.

Pegues DA, Ohl ME, Miller SI. 2005. *Salmonella* species, including *Salmonella* Typhi. Pages 2636–2654 in Principles and Practice of Infectious Diseases No. 2. G. L. Mandell, J. E. Bennett. and R. Dolin, ed. Elsevier Churchill Livingstone, Philadelphia, PA.

Peters TM, Maguire C, Threlfall EJ, Fisher IS, Gill N, Gatto AJ. 2003. The Salm-gene project—a European collaboration for DNA fingerprinting. *Euro. Surveill.* **8**: 46–50.

Phillips D, Jordan D, Morris S, Jenson I, Sumner J. 2008. A national survey of the microbiological quality of retail raw meats in Australia. *J. Food Prot.* **71**: 1232–1236.

Poelma PL, Andrewes WH, Silliker JH. 1984. *Salmonella.* In M. L. Speck (ed.). Compendium of methods for the microbiological examination of foods, 2nd ed., American Public Health Association, Washington D.C., USA. 286–326.

Pohl P, Lintermans B, Mainil J, Deprez P. 1989. Production de vérocytotoxine par les *Escherichia coli* du porc. *Ann. Méd. Vét.* **133**: 31–38.

Popoff MY, Bockemuhl J, Gheesling LL. 2004. Supplement 2002 (no. 46) to the Kauffmann-White scheme. *Res Microbiol* **155**: 568-570.

Pradel N, Livrelli V, De Champs CD, Palcoux JB, Reynaud A, Scheutz F, Sirot J, Joly B, Forestier C. 2000. Prevalence and characterization of Shiga toxin-producing *Escherichia coli* isolated from cattle, food, and children during a one-year prospective study in France. *J. Clin. Microbiol.* **38**: 1023–1031.

Prescott JF. 2000. Antimicrobial drug resistance and its epidemiology. In: Prescott JF, Baggot JD and Walker RD. (Eds.), Antimicrobial Therapy in Veterinary Medicine. Iowa State University Press, Ames. 27–49.

Price JF, Schweigert B. 1971. The Service of Meat and Meat Products 2nd Ed., W. H. Freeman and Company, San Francisco, USA.

PulseNet. 2002. One-day (24–48 h) standardized laboratory protocol for molecular subtyping of *Escherichia coli* O157:H7, non-typhoidal Salmonella serotypes, and Shigella sonnei by pulsed field gel electrophoresis (PFGE). Available at: http://www.cdc.gov/pulsenet/protocols/ecoli-salmonella-shigella-protocols.pdf. Accessed 11 July 2006.

Qadri F, Ann-Mari Svennerholm AM, Faruque ASG, Sack RB. 2005. Enterotoxigenic *Escherichia coli* in developing countries: epidemiology, microbiology, clinical features, treatment, and prevention. *Clin. Microbiol. Rev.* **18**: 465–483.

Raji MA, Minga UM, Machang'u RS. 2006. Prevalence and characterization of verotoxigenic *Escherichia coli* O157 isolated from local chickens in Morogoro, Tanzania. *J. Anim. Vet. Adv.* **5**: 952–958.

Randell AW. 2002. Scientific basis for setting food Standard through Codex Alimentarius. *Food Nutr. Agric.* **31**: 7-11.

Raufu I, Hendriksen RS, Ameh JA, Aarestrup FM. 2009. Occurrence and characterization of *Salmonella* Hiduddify from chickens and poultry meat in Nigeria. *Foodborne Pathog. Dis.* **6**: 425–430.

Reeves MW, Evins GM, Heiba AA, Pleikeytis BD, Farmer JJ. 1989. Clonal nature of *Salmonella typhi* and its genetic relatedness to other Salmonellae as shown by multilocus enzyme electrophoresis and proposal of Salmonella bongori comb.nov. *J. Clin. Microbiol.* **27**: 313–320.

Rhoades JR, Duffy G, Koutsoumanis K. 2009. Prevalence and concentration of verocytotoxigenic *Escherichia coli*, *Salmonella enterica* and *Listeria monocytogenes* in the beef production chain: a review. *Int. J. Food Microbiol.* **26**:357–376.

Rigobelo EC, Santo E, Marin JM. 2008. Beef carcass contamination by shiga toxin–producing *Escherichia coli* strains in an abattoir in Brazil: characterization and resistance to antimicrobial drugs. *Foodborne Pathog. Dis.* **5**: 811–817.

Riley LW, Remis RS, Helgerson SD, McGee HB, Wells JB, Davis R, Hebert RJ, Olcott ES, Johnson LM, Hargrett NT, Blake PA, Cohen ML. 1983. Hemorrhagic colitis associated with a rare *Escherichia coli* serotype. *New Engl. J. Med.* **308**: 681–685.

Robin AC, Ziprin Ri L. 2001: Bacteriology of *Salmonella* U.S. Department of Agriculture. College Stcrtioiz, Te.ucrs Copyright 0 by Marcel Dekker, Inc.

Roels TH, Frazak PA, Kazmierczak JJ, Mackenzie WR, Proctor ME, Kurzynski TA, Davis JP. 1997. Incomplete sanitation of meat grinder and ingestion of raw ground beef: contributing factors to a large outbreak of *Salmonella* Typhimurium infection. *Epidemiol. Infect.* **119**: 127–134.

Rogerie F, Marecat A, Gambade S, Dupond F, Beaubois P, Lange M. 2001. Characterization of Shiga toxin producing *E. coli* and O157 serotype *E. coli* isolated in France from healthy domestic cattle. *Food Microbiol.* **63**: 217–223.

Rose N, Beaudeau F, Drouin P, Toux JY, Rose V, Colin P. 1999. Risk factors for *Salmonella enterica* subsp. *enterica* contamination in French broiler-chicken flocks at the end of the rearing period. *Prev. Vet. Med.* **39**: 265–277.

Rossolini GM, D'Andrea MM, Mugnaioli C. 2008. The spread of CTX-Mtype extended-spectrum beta-lactamases. *Clin. Microbiol. Infect.* **1**: 33–41.

Rúgeles LC, Bai J, Martínez AJ, Vanegas MC, Gómez-Duarte OG. 2010. Molecular characterization of diarrheagenic *Escherichia coli* strains from stools samples and food products in Colombia. *Int. J. Food Microbiol.* **138**: 282–286.

Rwego IB, Gillespie TR, Isabirye-Basuta G, Goldberg TL. 2008. High rates of *Escherichia coli* transmission between livestock and humans in rural Uganda. *J. Clin. Microbiol.* **46**: 3187–3191.

Rycroft AN. 2000. Structure, function and synthesis of surface polysacharids in *Salmonella* in domestic animals. CAB international, eds. Wray,C. and Wray, A. 19–33.

Sackey BA, Mensah P, Collison E, Sakyi-Dawson E. 2000. *Campylobacter, Salmonella, Shigella* and *Escherichia coli* in live and dressed poultry from metropolitan Accra. *Int. J. Food Microbiol.* **71**: 21–28.

Salyers AA, Gupta A, Wang Y. 2004. Human intestinal bacteria as reservoirs for antibiotic resistance genes. *Trends Microbiol.* **12**: 412–416.

Sansonetti P, Zychlinsky A. 2002. Methods in Microbiology Volume 31: Molecular cellular Microbiology Academic Press International.

Sartz L, De Jong B, Hjertqvist M, Plym-Forshell L, Alsterlund R, Löfdahl S, Osterman B, Ståhl A, Eriksson E, Hansson HB, Karpman D. 2008. An outbreak of *Escherichia coli* O157:H7 infection in southern Sweden associated with consumption of fermented sausage; aspects of sausage production that increase the risk of contamination. *Epidemiol Infect.* **136**: 370-380.

Schimmer B, Nygard K, Eriksen HM, Lassen J, Lindstedt BA, Brandal LT, Kapperud G, Aavitsland P. 2008. Outbreak of haemolytic uraemic syndrome in Norway caused by stx2-positive *Escherichia coli* O103:H25 traced to cured mutton sausages. *BMC Infect. Dis.* **8**: 41.

Schmidt H, Beutin L, Karch H. 1995. Molecular analysis of the plasmid-encoded hemolysin of *Escherichia coli* O157:H7 strain EDL 933. *Infect. Immun.* **63**: 1055–1061.

Schmidt H, Geitz C, Phillips IT, Mathias F, Karch H. 1999. Non-O157 pathogenic Shiga toxin producing *Escherichia coli*: Phenotypic and genetic profiling of virulence traits and evidence for clonality. *J. Infect. Dis.* **179**:115–123.

Schmidt MA. 2010. LEEways: tales of EPEC, ATEC and EHEC. *Cell. Microbiol.* **12**: 1544–1552.

Schouten JM, van de Giessen AW, Frankena K, De Jong MC, Graat EA. 2005. *Escherichia coli* O157 prevalence in Dutch poultry, pig finishing and veal herds and risk factors in Dutch veal herds. *Prev. Vet. Med.* **70**: 1–15.

Schwartz DC, Cantor CR. 1984. Separation of yeast chromosome-sized DNAs by pulsed field gradient gel electrophoresis. *Cell.* **37**: 67-75.

Schwarz S, Kehrenberg C, Walsh TR. 2001. Use of antimicrobial agents in veterinary medicine and food animal production. *Int J Antimicrob Agents* **17**: 431–437.

Service des Statistiques Animales. 2007. Les statistiques de l'élevage au Burkina Faso. Ministère des Ressources Animales. Sécrétariat Général. 75p.

Shelobolina ES, Sullivan SA, O'Neill KR, Nevin KP, Lovley DR. 2004. Isolation, characterization, and U(VI)-reducing potential of a facultatively anaerobic, acid-resistant Bacterium from Low-pH, nitrate- and U(VI)-contaminated subsurface sediment and description of *Salmonella subterranea* sp. nov. *Appl Environ Microbiol.* **70**: 2959–2965.

Sidjabat-Tambunan H, Bensink JC. 1997. Verotoxin-producing *Escherichia coli* from the faeces of sheep, calves and pigs. *Aust. Vet. J.* **75**: 292–293.

Simporé J, Ouermi D, Ilboudo D, Kabre A, Zeba B, Pietra V, Pignatelli S, Nikiema JB, Kabre GB, Caligaris S, Schumacher F, Castelli F. 2009. Aetiology of acute gastro-enteritis in children at Saint Camille Medical Centre, Ouagadougou, Burkina Faso. *Pakist. J. Biol. Sci.* **12**: 258–263.

Singer RS, Mayer AE, Hanson TE, Isaacson RE. 2009. Do microbial interactions and cultivation media decrease the accuracy of *Salmonella* surveillance systems and outbreak investigations? *J. Food Prot.* **72**: 707–713.

SkermanVBD, Mc GowanV, Sneath PHA. 1980. Approved list of bacterial names. *Int. J. Syst. Bacterial.* **30**: 225–420.

Sofos JN, Kochevar SL, Bellinger GR, Buege DR, Hancock DD, Ingham SC, Morgan JB, Reagan JO, Smith GC. 1999. Source and extent of microbiological contamination of beef carcasses in seven United States slaughtering plants. *J. Food Prot.* **62**: 140–145.

Sperandio V, Kaper JB, Bortolini MR, Neves BC, Keller R, Trabulsi LR. 1998. Characterization of the locus of enterocyte effacement (LEE) in different enteropathogenic *Escherichia coli* (EPEC) and Shiga-toxin producing *Escherichia coli* (STEC) serotypes. *FEMS Microbiol. Lett.* **164**: 133–139.

Stern NJ, Fedorka-Cray P, Bailey JS, Cox NA, Craven SE, Hiett KL, Musgrove MT, Ladely S, Cosby D, Mead. GC. 2001. Distribution of *Campylobacter* spp. in selected U.S. poultry production and processing operations. *J. Food Prot.* **64**: 1705–1710.

Stevens A, Kaboré Y, Perrier-Gros-Claude JD, Millemann Y, Brisabois A, Catteau M, Cavin JF, Dufour B. 2006. Prevalence and antibiotic-resistance of *Salmonella* isolated from beef sampled from the slaughterhouse and from retailers in Dakar (Senegal). *Int. J. Food Microbiol.* **110**: 178–186.

Stopforth JD, Lopes M, Shultz JE, Miksch RR, Samadpour M. 2006. Location of bung bagging during beef slaughter influences the potential for spreading pathogen contamination on beef carcasses. *J Food Prot.* **69**: 1452-1455.

Strockbine NA, Wells JG, Bopp CA, Barrett TJ. 1998. Overview of detection and subtyping methods, p 331-356. In: Kaper JB, O'Brien AD, eds. *Escherichia coli* O157:H7 and other Shiga toxin producing *E. coli* strains. American Society for Microbiology, Washington, DC.

Sullivan A, Edlund C, Nord CE. 2001. Effect of antimicrobial agents on the ecological balance of human microflora. *Lancet Infect. Dis.* **1**: 101–114.

Sussman JK, Simons EL, Simons RW. 1996. *Escherichia coli* translation initiation factor 3 discriminates the initiation codon in vivo. *Mol. Microbiol.* **12**: 347–60.

Swerlow, Altekruse. 1998. Emerg. Infect. 2, eds. Scheld,W.M. et col. ASM Press,Washington D.C. 273–294.

Tarr PI, Gordon CA, Chandler WL. 2005. Shiga-toxin-producing *Escherichia coli* and haemolytic uraemic syndrom. *Lancet* **365**: 1073–1096.

Thomason BM, Biddle JW, Cherry WB. 1975. Detection of salmonellae in the environment. *Appl. Microbiol.* **30**: 764–767.

Thorns CJ. 2000. Bacterial food-borne zoonoses. *Rev Sci Tech.* **19**:226–239.

Threlfall EJ. 2002. Antimicrobial drug resistance in *Salmonella*: problems and perspectives in food- and water-borne infections. *FEMS Microbiol. Rev.* **26**: 141–148.

Tibaijuka B, Molla B, Hildebrandt G, Kleer J. 2003. Occurrence of *Salmonella* in retail raw chicken products in Ethiopia. *Berl Munch Tierarztl Wochenschr.* **116**: 55–58.

Tilden J Jr, Young W, McNamara AM, Custer C, Boesel B, Lambert-Fair MA, Majkowski J, Vugia D, Werner SB, Hollingsworth J, Morris JG Jr.1996. A new route of transmission for *Escherichia coli*: infection from dry fermented salami. *Am. J. Public Health*. **86**: 1142–1145.

Todd EC. 1997. Epidemiology of foodborne diseases: a worldwide review. *World Health Stat.* **50**: 30–50.

Trabulsi LR, Keller R, Tardelli Gomes TA. 2002. Typical and atypical enteropathogenic *Escherichia coli*. *Emerg. Infect.Dis.* **8**: 508–513.

Tuttle J, Gomez T, Doyle MP, Wells JG, Zhao T, Tauxe RV, Griffin PM. 1999. Lessons from a large outbreak of *Escherichia coli* O157:H7 infections: insights into the infectious dose and method of widespread contamination of hamburger patties. *Epidemiol. Infect.* **122**: 185–197.

Uber AP, Trabulsi LR, Irino K, Beutin L, Ghilardi ÅCR, Gomes TAT, Liberatore AMA, de Castro AFP, Elias WP. 2006. Enteroaggregative *Escherichia coli* from humans and animals differ in major phenotypical traits and virulence genes. *FEMS Microbiol. Lett.* **256**: 251–257.

Umoh VJ, Odoba MB. 1999. Safety and quality evaluation of street foods sold in Zaria, Nigeria. *Food Contr.* **10**: 9–14.

Ungemach FR, Müller-Bahrdt D, Abraham G. 2006. Guidelines for prudent use of antimicrobials and their implications on antibiotic usage in veterinary medicine. *Inter. J. Med. Microbiol.* **296**: 33–38.

Valdezate S, Vidal A, Herrera-León S, Pozo J, Rubio P, Usera MA, Carvajal A, Echeita MA. 2005. *Salmonella* Derby clonal spread from pork. *Emerg. Infect. Dis.* 11: 694–698.

Van den Bogaard AE, Stobberingh EE. 2000. Epidemiology of resistance to antibiotics. Links between animals and humans. *Int. J. Antimicrob. Agents* 14: 327–335.

Van Nierop W, Dusé A G, Marais E, Aithma N, Thothobolo N, Kassel M, Stewart R, Potgieter A, Fernandes B, Galpin JS, Bloomfield SF. 2005. Contamination of chicken carcasses in Gauteng, South Africa, by *Salmonella*, *Listeria monocytogenes* and *Campylobacter*. *Int. J. Food Microbiol.* 99: 1–6.

Varma JK, Molbak K, Barrett TJ, Beebe JL, Jones TF, Rabatsky-Ehr T, Smith KE, Vugia DJ, Chang HG, Angulo FJ. 2005. Antimicrobial-resistant nontyphoidal *Salmonella* is associated with excess bloodstream infections and hospitalizations. *J Infect Dis.* 191:554–561.

Vedantam G, Hecht DW. 2003. Antibiotics and anaerobes of gut origin. Curr. *Opin. Microbiol.* 6: 457–461.

Vidal M, Kruger E, Durán C, Lagos R, Levine M, Prado V, Toro C, Vidal R. 2005. Single multiplex PCR assay to identify simultaneously the six categories of diarrheagenic *Escherichia coli* associated with enteric infections. *J. Clin. Microbiol.* 43: 5362–5365.

Vollaard AM, Ali S, van Asten HA, Ismid IS, Widjaja S, Visser LG, Surjad Ch, van Dissel JT. 2004. Risk factors for transmission of foodborne illness in restaurants and street vendors in Jakarta, Indonesia. *Epidemiol. Infect.* 132: 863–872.

Vorster SM, Greebe RP, Nortje GL. 1994. Incidence of *Staphylococcus aureus* and *Escherichia coli* in ground beef, broilers and processed meats in Pretoria, South Africa. *J. Food Prot.* 57: 305–310.

Wayne LG, Brenner DG, Colwell RR, Grimont PAD, Kandler O, Krichevsky MI, Moore H, Moore WEC, Murray RGE, Stackbrandt E, Starr MP,

Trüper HG.1987. Report of the ad-hoc commitee on reconciliation of approaches to bacterial systemics. *Int. J. Syst. Bacteriol.* **37**: 463–464.

Weill FX. 2006. Synthèse des résultats du CNR : Souches humaines, 2005. 10ème reunion annuelle du réseau *Salmonella*. Afssa, Maisons Alfort, Paris.

White DG, Zhao S, Sudler R, Ayers S, Friedman S, Chen S, McDermott PF, McDermott S, Wagner DD, Meng J. 2001. The isolation of antibiotic resistant *Salmonella* from retail ground meat. *New Engl. J. Med.* **345**: 1147–1154.

White DG, Zhao S, Sulder R, Ayers S, Friedman S, Chen S, McDermott PF, McDermott S, Wagner DD, Meng J. 2001. The isolation of antibiotic-resistant *Salmonella* from retail ground meat. *New Engl. J. Med.* **345**:1147–1154.

White PL, Baker AR, James WO. 1997: Strategies to control *Salmonella* and *Campylobacter* in raw poultry products. *Rev. Sci. Tech.* **16**: 525–541.

Whittam TS, Wolfe ML, Wachsmuth IK, Orskov F, Orskov I, Wilson RA. 1993. Clonal relationships among *Escherichia coli* strains that cause hemorrhagic colitis and infantile diarrhea. *Infect. Immun.* **61**: 1619–1629.

Wieler LH, McDaniel TK, Whittam TS, Kaper JB. 1997. Insertion site of the locus of enterocyte effacement in enteropathogenic and enterohemorrhagic *Escherichia coli* differs in relation to the clonal phylogeny of the strains. *FEMS Microbiol. Lett.* **156**: 49–53.

Wihelmi I, Roman E, Sanchez-Fauquier A. 2003. Virus causing gastroenteritis. *Clin. Microbiol. Infect.* **9**: 247–262.

Wilson IG. 2002. *Salmonella* and *Campylobacter* contamination of raw retail chickens from different producers: a six year survey. *Epidemiol Infect.* **129**: 635-645.

Winfield MD, Groisman EA. 2003. Role of nonhost environments in the lifestyles of *Salmonella* and *Escherichia coli*. *Appl. Environ. Microbiol.* **69**: 3687–3694.

Wong TL, Nicol C, Cook R, MacDiarmid S. 2007. *Salmonella* in uncooked retail meats in New Zealand. *J. Food Prot.* **70**: 1360–1365.

Woodward DL, Khakhria R, Johnson WM. 1997. Human salmonellosis associated with exotic pets. *J. Clin. Microbiol.* **35**: 2786–2790.

Wouafo M, Nzouankeu A, Kinfack JA, Fonkoua MC, Ejenguele G, Njine T, Ngandjio A. 2010. Prevalence and antimicrobial resistance of *Salmonella* serotypes in chickens from retail markets in Yaounde (Cameroon). *Microb. Drug. Resist.* **16**: 171–176.

Xia X, Meng J, McDermott PF, Ayers S, Blickenstaff K, Tran TT, Abbott J, Zheng J, Zhao S. 2010. Presence and characterization of shiga toxin-producing *Escherichia coli* and other potentially diarrheagenic *E. coli* strains in retail meats. *Appl. Environ. Microbiol.* **76**: 1709–1717.

Yan SS, Pendrak ML, Abela-Ridder B, Punderson JW, Fedorko DP, Foley SL. 2003. An overview of *Salmonella* typing public health perspectives. *Clin. Appl. Immunol. Rev.* **4**: 189–204.

Zhao C, Ge B, De Villena J, Sudler R, Yeh E, Zhao S, White DG, Wagner D, Meng J. 2001. Prevalence of *Campylobacter* spp, *Escherichia coli*, and *Salmonella* serovars in retail chicken, turkey, pork, and beef from the Greater Washington, D.C., area. *Appl. Environ. Microbiol.* **67**: 5431–5436.

ANNEXES

ANNEXES

Fiche d'enquêtes

➢ **Etat de la boucherie**

Infrastructure à part entière	non/__/
oui/__/	
Installation d'eau et de sanitaire	non/__/
oui/__/	
Hangar en paillote et table de vente	non/__/
oui/__/	
Présence de poubelle	non/__/
oui/__/	
Présence de déchets solides	non/__/
oui/__/	
Présence d'eau stagnante	non/__/
oui/__/	
Atmosphère poussiéreuse	non/__/
oui/__/	
Présence d'animaux errants	non/__/
oui/__/	

➢ **Traitement subi aux viandes et autres**

Protection pendant le transport, vente et entreposage

non/__/   oui/__/

Transport des viandes

Véhicule frigorifique	non/__/
oui/__/	
Véhicule non frigorifique	non/__/
oui/__/	
Motocyclette	non/__/
oui/__/	

Bicyclette   non/__/   oui/__/

Mauvais conditionnement pendant la vente   non/__/
oui/__/

Mauvaise manipulation des viandes   non/__/
oui/__/

Viandes et abats vendue dans la même zone   non/__/
oui/__/

Viandes et abats vendues sur les mêmes tables   non/__/
oui/__/

Ustensile et autres matériels

Etat des tables de vente   sales/__/
propres/__/

Nettoyage des couteaux et tables avant vente   non/__/
oui/__/

Eau de lavage des mains   non/__/
oui/__/

Utilisation de savon   non/__/
oui/__/

➢ **Bouchers**

Sexe   M/__/   F/__/

Tranche d'âge   non/__/
oui/__/

Savoir lire et écrire   non/__/   oui/__/

Protection des cheveux   non/__/
oui/__/

Utilisation de blouse   non/__/
oui/__/

Propreté des vêtements   non/__/
oui/__/

Formation en hygiène   non/___/
oui/___/
Présence de torchons pour les mains   non/___/
oui/___/
Lavage des mains après les toilettes   non/___/
oui/___/

Annexes 2

Quelques exemples de structure antigéniques des *Salmonella* sp

Tableau IV: Les antisérums O et quelques exemples de structure antigéniques

Serums	O-antigen	Antigen structure	Controls strains
OA	2	1,2,12:a:1,5	S. Paratyphi A
(OMA)	4	4,12:i:1,2	S. Typhimurium var copenh.
	4,5	4,5,12:i:1,2	S. Typhimurium
	9	9,12:g,m:-	S. Enteritidis
	10	3,10,26:l,v:1,6	S. London
	15,34	3,15,34:e,h:1,6	S. Anatum var 15*,34*
	19	1,3,19:g,s,t:-	S. Senftenberg
	21	21:b:-	S. ssp I Ryhmä L
	27	1,4,12,27:l,v:1,7	S. Bredeney
	9,46	9,46:d:1,7	S. Strasbourg
OB	7	6,7:k:1,5	S. Thompson
(OMB)	8	6,8:e,h:1,2	S. Newport
	11	11:i:1,2	S. Aberdeen
	14	6,14:y:1,7	S. Carrau
	8,20	8,20:i:z6	S. Kentucky
	22	13,22:z:1,6	S. Poona
	23	13,23:d:1,7	S. Grumpensis
	24	6,14,24:e,h:1,5	S. Bahrenfeld
	25	1,6,14,25:y:1,7	S. Madelia

OMC	16	16:b:e,n,x	S. Hvittingfoss
	17	17:b:1,2	S. Kirkee
	18	18:z4,z23:-	S. Cerrro
	28	28:y:e,n,z15	S. Telaviv
	30	30:b:e,n,x	S. Urbana
	35	35:k:1,6	S. Adelaide
	38	38:k:1,6	S. Inverness
OMD	39	39:k:1,5	S. Champaign
	40	40:b:1,5	S. Riogrande
	41	41:z4,z23:-	S. Waycross
	42	42:z36:-	S. Weslaco
	43	43:f,g:-	S. Milwaukee
	44	44:a:l,w	S. Niarembe
	45	45:c:e,n,x	S. Deversoir
OME	47	47:i:e,n,z15	S. Bergen
	48	48:z52:z	S. ssp IIIb Ryhmä Y
	50	50:z:e,n,x	S. ssp II Ryhmä Z
	51	1,51:z:1,6	S. Treforest
	52	52:d:1,5	S. Utrecht
	53	53:z4,z24:-	S. ssp II Ryhmä O:53

Tableau IV: les antisérums O et quelques exemples de structure antigéniques (suite)

	61	61:i:z	S. ssp IIIb Ryhmä O:61
OMF	**54**	54:g,s,t:-	S. Uccle
	55	55:k:z39	S. ssp II Ryhmä O:55
	56	56:b:-	S. ssp II Ryhmä O:56
	57	57:z29:z42	S. ssp II Ryhmä O:57
	58	58:l,z13,z28:1,5	S. ssp II Ryhmä O:58
	59	59:k:(z)	S. ssp II Ryhmä O:59
OMG	**60**	60:z:e,n,x	S. ssp II Ryhmä O:60
	62	62:g,z57:-	S. ssp IIIa Ryhmä O:62
	63	63:z4,z32:-	S. ssp IIIa Ryhmä O:63
	64	48,64:i:z35,z57	S. ssp IIIb Ryhmä Y
	65	65:l,v:z	S. ssp IIIb Ryhmä O:65
	66	66:z35:-	S. ssp V Ryhmä O:66
	67	67:r:1,2	S. Crossness
Vi	**Vi**	9,12,Vi:d:-	S. Typhi

Tableau V : les antisérums H et quelques exemples de structure antigénique

Serums	H-antigen	Antigen structure	Controls strains
HA	a	47:a:-	S. ssp II Ryhmä X
(HMA)	b	56:b:-	S. ssp II Ryhmä O:56
	b	4,5,12:b:1,2	S. Paratyphi B
	c	35:c:-	S. Yolo
	d	38:d:-	S. ssp II Ryhmä P
	i	4,5,12:i:1,2	S. Typhimurium
	i	8,20:i:-	S. ssp I Ryhmä C3
	z10	3,15,34:z10:-	S. ssp I Ryhmä E1
	z29	6,7:z29:-	S. Tennesee
HB	e,h	39:e,h:-	S. ssp I Ryhmä Q
(HMB)	e,n,x	4,12:-:e,n,x	S. Abortusequi
	e,n,z15	4,12:e,n,z15:-	S. ssp I ryhmä B
	f,g	4,12:f,g:-	S. Derby
	g,m,s	6,7:g,m,s:-	S. Montevideo

Tableau V: Les antisérums H et quelques exemples de structure antigénique (suite)

	g,m,q	9,12:g,m,q:-	S. Blegdam
	g,p	1,9,12:g,p:-	S. Dublin
	g,q	9,12:g,q:-	S. Moscow
	g,p,u	1,9,12:g,p,u	S. Rostock
	m,t	6,7:m,t:-	S. Oranienburg
HC	**k**	6,7:k:-	S. ssp I Ryhmä C1
(HMC)	**l,v**	16:l,v:-	S. ssp I Ryhmä I
	l,w	13,23:-:lw	S. ssp I Ryhmä G
	l,z13	3,10:l,z13:1,5	S. Uganda
	l,z28	1,9,12:l,z28:-	S. sspI Ryhmä G
	r	42:r:-	S. ssp II Ryhmä T
	y	28:y:-	S. ssp I Ryhmä M
	z	13,23:z:-	S. ssp I Ryhmä G
	z4,z23	18:z4,z23:-	S. Cerro
	z24	53:z4,z24:-	S. ssp II Ryhmä O:53
HD	**z35**	4,12:z35:-	S. ssp I Ryhmä B
(HMD)	**z36**	18:z36:-	S. ssp II Ryhmä K
	z38	43:z38:-	S. Irigny
	z39	40:a:z39	S. ssp II Ryhmä R
	z41	40:z41:1,2	S. Karamoja
	z42	6,7:b:z42	S. ssp II Ryhmä C1
	z44	47:z44:-	S. Quinhon
HE	**1,2**	4,5,12:-:1,2	S. ssp I Ryhmä B
(HME)	**1,5**	6,7:-;1,5	S. ssp I Ryhmä C1
	1,6	9,12:-:1,6	S. ssp II Ryhmä D

	1,5,7	52:-:1,5,7	S. ssp II Ryhmä O:52
	z6	8,20:i:z6	S. Kentucky
HF	**z52**	50:z52:z35	S. ssp IIIb Ryhmä Z
(HMF)	**z53**	60:(k):z53	S. ssp IIIb
	z54		
	z55		
	z57	48,64:i:z35,z57	S. ssp IIIb Ryhmä Y
Hz29	**z29**	6,7:z29:-	S. Tennesee
Hz51	**z51**	30:g,z51:-	S. Wayne
Hz60	**z60**	13,23:z:z60	S. ssp I
Hz61	**z61**		
Hz64	**z64**	18:z4,z23:z64	S. Aarhus
Hz65	**z65**	59:k:z65	S. ssp II Ryhmä O:59
Hz67	**z67**		
Hz68	**z68**		
Hz69	**z69**		
Hz81	**z81**	48:z81:-	S. bongori

Annexe 3

Poster 1: Common occurrence of various STEC serotypes in meats sold at local markets in Ouagadougou, Burkina Faso, Presenté au FEMS, 2011

Common occurrence of various STEC serotypes in meats sold at local markets in Ouagadougou, Burkina Faso

Haukka K[1], Kagambega A[2], Martikainen O[1], Traore AS[2], Barro N[2] & Siitonen A[1]

[1] Bacteriology Unit, Department of Infectious Disease Surveillance and Control, National Institute for Health and Welfare, POB 30, 00271 Helsinki, Finland, [2] Laboratoire de Biochimie et de Biologie Moléculaire, Centre de Recherche en Sciences Biologiques, Alimentaires et Nutritionnelles, Département de Biochimie-Microbiologie, UFR-SVT-Université de Ouagadougou, 03 B.P. 7021 Ouagadougou 03 ; Burkina Faso

INTRODUCTION

Infection with Shiga toxin-producing *E. coli* (STEC) can cause serious intestinal infection, haemolytic-uremic syndrome and permanent renal injuries especially in children. Also other pathogroups of *E.coli*, namely enteropathogenic *E. coli* (EPEC), enterotoxigenic *E. coli* (ETEC), enteroinvasive *E. coli* (EIEC) and enteroaggregative *E. coli* (EAEC) can cause diarrhea. However, little is known on the occurrence of these pathogens in developing countries. Therefore, our objective was to reveal the prevalence of the 5 *E. coli* pathogroups in the meats sold for human consumption at common markets in Burkina Faso. Furthermore, we isolated STEC strains in order to characterize them further.

METHODS

We assessed the microbial quality of raw meat sold on 4 open markets in Ouagadougou. A total of 120 beef meat, beef intestine, mutton and chicken samples were collected. 120 colonies with typical *E. coli* morphology growing on EMB agar and 120 mixed cultures growing on MH agar were subjected to 16-plex PCR amplifying 16 genes specific for STEC, EPEC, ETEC, EIEC and EAEC [1]. The tested genes were *uidA, pic, bfp, invE, hlyA, elt, ent, escV, eaeA, ipaH, aggR, stx1, stx2, estIa, estIb* and *ast*. STEC positive mixed cultures were subjected to colony hybridization using *stx1* and *stx2* probes, in order to isolate STEC strains. The obtained isolates were serotyped by traditional methods and genotyped by PFGE.

RESULTS

16-plex PCR was used to detect 5 diarrheagenic *E. coli* pathogroups (Fig. 1). *E. coli* were detected in 51 (43%) of the 120 meat samples; in 16 (44%) beef, in 19 (53%) beef intestine, in 9 (37%) mutton and in 7 (29%) chicken samples. The pathogens detected included 33 STECs (28%), 14 EPECs (12%), 10 ETECs (8%) and 5 EAECs (4%) (Table 1). No EIEC virulence determinants were detected in any of the meat types. In several samples virulence determinants for two pathogroups were observed (Table 1) but determination of the presence of STEC+EPEC was not possible because STEC strains can contain all the virulence genes present in atypical EPEC.

The 20 STEC isolates represented 14 different serotypes (Table 2). Many of the serotypes have been previously isolated from humans with diarrhea. PFGE further separated some of the isolates with the same serotype from each other.

Table 2. Serotypes of the 20 STEC isolates and their main virulence genes detected by PCR.

Strain Number	Sample	Serotype	stx1	stx2	EHEC-hly
FE 95160	Intestine	O2:H2	+	-	+
FE 95562	Beef	O2:H2	+	-	+
FE 94914	Beef	O8:H-	+	+	+
FE 95928	Intestine	O20:H16	+	-	-
FE 95385	Intestine	O38:H26	+	+	+
FE 95061	Beef	O43:H2	+	+	-
FE 95614	Beef	O43:H2	+	+	-
FE 95260	Beef	O43:H2	+	-	+
FE 95381	Intestine	O74:H42	+	+	+
FE 95433	Mutton	O76:H-	+	+	+
FE 96003	Intestine	O77:H27	+	-	-
FE 94898	Intestine	O82:H8	+	+	+
FE 95201	Beef	O82:H8	+	+	+
FE 95612	Mutton	O126:H8	+	-	+
FE 95105	Beef	O139:H25	+	-	-
FE 95104	Intestine	O139:H25	+	-	-
FE 94765	Mutton	O174:H8	+	-	-
FE 95383	Intestine	O179:H-	+	+	+
FE 95258	Intestine	OX183:H18	+	-	+
FE 95610	Beef	OX183:H18	+	+	+

Table 1. Prevalence of pathogenic *E. coli* (%).

E. coli pathogroups	Beef n=36	Intestine n=36	Mutton n=24	Chicken n=24	Total n=120
STEC	12 (33%)	16 (44%)	5 (21%)	0	33 (28%)
EPEC	4 (11%)	0	3 (13%)	7 (29%)	14 (12%)
ETEC	2 (6%)	6 (17%)	1 (4%)	1 (4%)	10 (8%)
EAEC	1 (3%)	3 (8%)	1 (4%)	0	5 (4%)
STEC+ETEC	2(6%)	3 (8%)	1 (4%)	0	6 (5%)
STEC+EAEC	1(3%)	3 (8%)	0	0	4 (3%)
EPEC+ ETEC	0	0	0	1 (4%)	1 (1%)

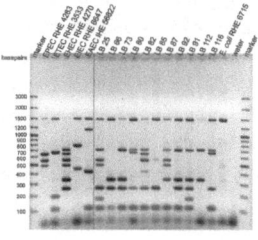

Fig.1. 16-plex PCR result from the meat samples.

CONCLUSION

Our work demonstrates that raw meats sold at open markets in Ouagadougou commonly contain *E. coli* pathogroups, especially STEC. These data are valuable for a risk assessment for foodborne illnesses linked to raw meat and emphasize the importance of improved hygiene in handling of meat.

Reference

[1] Antikainen *et al.* Eur. J. Clin. Microbiol. Inf. Diseases 28, 899–908, 2009.

Acknowledgements

The work was funded by theThe study was funded by the Academy of Finland grant 122600 to collaboration between the Finnish National Institute for Health and Welfare (THL) and CRSBAN/University of Ouagadougou.

Terveyden ja hyvinvoinnin laitos • Institutet för hälsa och välfärd • National Institute for Health and Welfare

Abstract for oral presentation IMMEM -9

Detection and characterization of EHEC isolates from meat samples collected from local markets in Ouagadougou, Burkina Faso

Haukka K[1], Kagambega A[2], Martikainen O[1], Barro N[2], Traore AS[2] & Siitonen A[1]

[1] *Bacteriology Unit, Department of Infectious Disease Surveillance and Control, National Institute for Health and Welfare, POB 30, 00271 Helsinki, Finland*

[2] *Département de Biochimie-Microbiologie, UFR-SVT-Université de Ouagadougou, 03 B.P. 7021 Ouagadougou 03 , Burkina Faso*

Microbiological safety of food stuffs sold on common markets in developing countries has been poorly studied. In our joint project, we try to reveal the occurrence and transfer of the enteropathogenic bacteria from the food and environmental sources to humans in Burkina Faso. In the current study, we assessed the microbial quality of raw meat sold on 4 open markets in Ouagadougou. A total of 120 chicken, mutton, beef meat and beef intestine samples were collected and cultured for diarrheagenic *E. coli* strains, among others. 120 colonies with typical *E. coli* morphology growing on EMB agar and 120 mixed cultures growing on MH agar were subjected to 16-plex PCR amplifying 16 genes specific for EPEC, ETEC, EHEC, EAEC or EIEC in a single PCR reaction. 7 of the 120 single colonies (6%) were EHECs and as many as 33 from the 120 mixed cultures (28%) were positive for EHEC. Mostly EHEC was detected in beef meat or intestines (in 36% of the beef samples), followed by mutton (in 29% of the mutton samples), never in chicken. Also pathotypes EPEC and ETEC were detected. EHEC positive mixed cultures were subjected to colony hybridization using *stx1* and *stx2* probes, in order to isolate pure cultures. This far we have isolated 15 EHEC strains using the probes. All of the 22 currently available EHEC strains are being serotyped using both real-time PCR for the most common human pathogenic serotypes and by traditional antisera. After serotyping, the strains will be subjected to genotypic characterization by PFGE. In conclusion, prevalence of EHEC in the meat sold on common markets in Ouagadougou was high. Infection with certain EHEC serotypes can cause serious symptoms and permanent renal injuries to children, but very little is known on their occurrence in the developing countries. This far, we have not isolated any EHEC from diarrheagenic children studied in our project. However, the common occurrence of EHEC shows that proper risk assessment for meat sold on markets is needed.

MIX

Papier | Fördert
gute Waldnutzung

FSC® C083411

Zeitfracht Medien GmbH
Ferdinand-Jühlke-Straße 7
99095 Erfurt, Deutschland
produktsicherheit@kolibri360.de

Druck:
CPI Druckdienstleistungen GmbH
im Auftrag der
Zeitfracht Medien GmbH
Ein Unternehmen der Zeitfracht - Gruppe
Ferdinand-Jühlke-Str. 7
99095 Erfurt